LEARNING AND INFORMATION
SERVICES
UNIVERSITY OF CUMBRIA

Housing Decisions in Later Life

Also by Roger Clough, Mary Leamy, Vince Miller and Les Bright

CARE IN CHAOS: Frustration and Challenge in Community Care R. Clough (*co-author with R. Hadley*)

GROUPS AND GROUPINGS: LIFE AND WORK IN DAY AND RESIDENTIAL CENTRES R. Clough (*co-author with A. Brown*)

HOMING IN ON HOUSING: A Study of Housing Decisions of People Aged over 60 R. Clough, M. Leamy, L. Bright, V. Miller L. Brooks

INSIGHTS INTO INSPECTION: THE REGULATION OF SOCIAL CARE R. Clough (*ed.*)

OLD AGE HOMES R. Clough

PRACTICE, POLITICS & POWER IN SOCIAL SERVICES DEPARTMENTS R. Clough

RESIDENTIAL WORK R. Clough

THE ABUSE OF CARE IN RESIDENTIAL INSTITUTIONS R. Clough (*ed.*)

THE PRACTICE OF RESIDENTIAL WORK R. Clough

Housing Decisions in Later Life

Roger Clough
Eskrigge Social Research, UK

Mary Leamy
Department of Psychology, University of Teesside, UK

Vince Miller
Department of Sociology, University of Kent, UK

and

Les Bright
Independent Consultant, UK

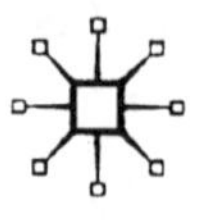

© Roger Clough, Mary Leamy,
Vince Miller and Les Bright 2004.

All rights reserved. No reproduction, copy or transmission of this publication may be made without written permission.

No paragraph of this publication may be reproduced, copied or transmitted save with written permission or in accordance with the provisions of the Copyright, Designs and Patents Act 1988, or under the terms of any licence permitting limited copying issued by the Copyright Licensing Agency, 90 Tottenham Court Road, London W1T 4LP.

Any person who does any unauthorized act in relation to this publication may be liable to criminal prosecution and civil claims for damages.

The authors have asserted their rights to be identified as the authors of this work in accordance with the Copyright, Designs and Patents Act 1988.

First published in 2004 by
PALGRAVE MACMILLAN
Houndmills, Basingstoke, Hampshire RG21 6XS and
175 Fifth Avenue, New York, N.Y. 10010
Companies and representatives throughout the world

PALGRAVE MACMILLAN is the global academic imprint of the Palgrave Macmillan division of St. Martin's Press, LLC and of Palgrave Macmillan Ltd. Macmillan® is a registered trademark in the United States, United Kingdom and other countries. Palgrave is a registered trademark in the European Union and other countries.

ISBN 1–4039–1287–4 hardback

This book is printed on paper suitable for recycling and made from fully managed and sustained forest sources.

A catalogue record for this book is available from the British Library.

Library of Congress Cataloging-in-Publication Data

Housing decisions in later life / Roger Clough ... [et al.].
p. cm.
Includes bibliographical references and index
ISBN 1–4039–1287–4
1. Older people – Housing – Great Britain – Decision making.
2. Housing – Resident satisfaction – Great Britain. I. Clough, Roger.

HD7287.92.G7H69 2004
363.5′946′0941—dc22 2004050896

10 9 8 7 6 5 4 3 2 1
13 12 11 10 09 08 07 06 05 04

Printed and bound in Great Britain by
Antony Rowe Ltd, Chippenham and Eastbourne

Contents

List of Tables

Acknowledgements

In this book we draw on findings from a three-year research study funded by the Community Fund, formerly the National Lottery Charities Board. We express our thanks to them and formally recognize that this work could not have taken place without their support. The two host institutions were the Department of Applied Social Science at Lancaster University and Counsel and Care, a national voluntary organization. We want to place on record our acknowledgement of the support we received from both organizations.

We gained a great deal from the advice that came from the members of the Research Advisory Group. They provided us with questions, challenges and advice and became an important focus for research team members in the planning and development of the research.

Particular individuals have played their part in the research. Liz Brooks, now Research and Development Associate at Counsel and Care, was the Project Officer for the research. She contributed to Chapter 3 with information on physical aspects of ageing and their impact on cognitive capacity. Keith Percy is Professor of Continuing Education at Lancaster University and, together with colleagues, was very supportive in the development of the research methods course which is referred to in Chapter 2. Marian McCraith ran creative writing sessions at the end of the project and some of the rich material in the book comes from older people's work with her.

Those of us who have been involved in writing the book are only too aware of the way in which the writing of the book, particularly in the latter stages as we pressed to get the material to publishers on the due date, has been a central feature of our lives. Our thanks for support and tolerance – which we hope they know – to those around us. Tomorrow we will be free of this labour of love, at least until the proofs arrive.

However, the most important acknowledgement is to the 1017 older people who played some part in the research, as interviewees and interviewers, questionnaire respondents, contributors of written housing stories and members of panels. Our thanks to them is immense. They have given of their time and their wisdom. We think that what they have had to say helps in the painting of a more complete picture of the housing journeys of older people. Insofar as that is true, it is their experiences which have enriched the findings.

Introduction

Research team members

Housing Decisions in Old Age, the study from which this book developed, was a partnership in research between *Counsel and Care*, a registered charity providing information and advice for older people, and campaigning for improvements in the quality of care, and the *Department of Applied Social Science* at *Lancaster University*. The staff of the research partnership were:

Les Bright Deputy Chief Executive, Counsel and Care, one day per week
Now Independent consultant, Bright Solutions

Liz Brooks Project Officer, Counsel and Care, half time
Now Research and Development Associate, Counsel and Care

Roger Clough Research Director, Professor of Social Care, Lancaster University, one day per week
Now Independent researcher, Eskrigge Social Research

Mary Leamy Senior Research Associate, Lancaster University, full time
Now Senior Lecturer in Psychology, Teesside University

Vince Miller Research Associate, Lancaster University, half time
Now Lecturer in Sociology, University of Kent

Members of the Research Advisory Group

Sue Adams, Director, Care and Repair England
Trevor Baker, formerly Research Manager, Hanover Housing
Bob Bessell, Director, Retirement Security Limited
Maria Evandrou, Senior Lecturer, Kings College, University of London
John Gatward, Chief Executive, Hanover Housing
Tracey Hylton, Research and Development Officer, Methodist Homes
Darshan Matharoo, Assistant Director (Investment), Housing Corporation
Angela McCullagh, Research Manager, The Gift of Thomas Pocklington
Steve Ongeri, Senior Policy Analyst, Housing Corporation

Robert Parkinson, formerly Policy and Information Officer, Methodist Homes
Daniel Pearson, Special Projects Manager, Help the Aged
Annie Stevenson, Care Policy Manager, Help the Aged

There is not space to list all the elders who participated in the research. However, we do list those who had the most involvement with the research development, the interviewers and panellists.

Interviewers

Lancaster students
John Blackburn
David Fox
Bert Green
Barbara Hawkes
Pat O'Connor
Mary Pallister
Gwyneth Raymond
George Steele
Chris Warren
Colin Watkins
Pam Wilson

London students
James Barrett
Carol Dapp
Mathiyaparanam Gnanasambanthan
Desmond Hall
Margaret Hanington
Allan Mitchell
Robert Neighbour
Heather Pierce
Sylvia Philpin-Jones

Panel members

Lambeth
Mr S. G. Bhatia
Mrs Toni Blackwood
Mrs Beryl Browne
Mrs Irma Critchlow
Mrs Irene Dike
Mr Basil Grant
Mrs Sheila Grant
Mr Des Hall
Mrs Gloria Planter
Mr Abdul Virani

Lancaster
Ms Edith Bell
Mr E. Cox
Mrs Joan Mollett
Mr Norman Ridley
Mr John Stoessl
Mrs Barbara Theobald
Mrs Brenda Thomas
Mrs Betty Valentine
Mrs Beth Wood

Leicester
Mr Walt Lindsay
Mr K. S. Sandhu
Mrs J. Gohil
Miss R. Waters
Mrs M. Izzard

Tyne and Wear
Mr John Berry
Mrs Mary Blackburn
Mrs Mary Craig
Mrs C. Galpin (deceased)
Mrs A. Macfarlane (deceased)
Ms Eleanor Dawson
Ms Irene Frisby
Mr Jim Harper
Mrs Millie Moore
Ms Stella Robinson
Mr Steve Ryan
Ms Pat Verbaan (deceased)
Mr Norman Wilson (deceased)

Weston-super-Mare
Mrs Rita Booth
Mrs J. Frossard
Mrs P. Hole
Mrs J. Humphries
Mrs M. Jones
Mrs Gwen Kiss
Mrs D. Morgan
Mrs Richards

In the book there are numerous quotations from written housing stories and interviews. All names used in such extracts, whether of individuals or housing organizations, are pseudonyms. Anonymity was guaranteed to participants and we have maintained this.

Research activity and the writing of this book

The research team was based in two places: Counsel and Care in London and the department of Applied Social Science at Lancaster. The team members have no doubt that the partnership between the two organizations, and the individuals within them, played a major part in the success of the project.

Counsel and Care with Les Bright as lead officer and Liz Brooks as project officer organized the panels in their five locations, the media releases throughout the project but of particular importance at the beginning when seeking written housing stories, and the holding of the budget and administration with the Community Fund. However, the importance of partnership lies in informal relationships as well as formal structures. So Counsel and Care were able to supply an advice line in case interviewers met people who wanted assistance and to support the London interviewers.

At Lancaster as Senior Research Associate and the only full-time member of the research team, Mary Leamy had the responsibility for the collection and analysis of the data which is drawn on extensively in the book. She was also the lead staff member in the development of the curriculum and the teaching of the Research Methods Course. Vince Miller worked with her on both the research methods course and the data collection and analysis. Roger Clough, as Research Director, had the overall responsibility for the project and, more specifically, worked with the Lancaster team.

Roger Clough and Mary Leamy have undertaken most of the planning and organization of the structure of the book, and have also taken the lead in the writing of the chapters. Chapters 2, 4, 5, 6 and 7 draw extensively on data from the research data written up by Mary Leamy and Vince Miller; they have been crafted by Mary Leamy with Vince Miller sharing the work on Chapters 5, 6 and 7. Roger Clough has worked on the other chapters, with Les Bright making contributions.

Such divisions do not do justice to the development of ideas between us, the stimulation of discussions, and our excitement in working on this book. We believe that intangible benefits have come from the partnership, giving added value to the individual and organizational elements. Meetings of the whole research team provided valuable opportunities to explore understandings of the material and work at the direction of the research. Indeed, those benefits have been significant factors in maintaining the energy for the project and in the continued collaboration resulting in this book.

We hope we have done justice to the trust of the older people who demanded that we make the case for changes in the understanding of their housing experiences and in the quality of the housing available in later life.

1
Houses and People: The Construction of Living Arrangements

The impact of housing

As we get older, the houses in which we live have a huge impact on our lives. In this book we try to get inside the experience of housing by examining the ways in which people construct, and re-construct, their lives in relation to their housing. In later life people are likely to spend longer inside their homes and, increasingly, to find some aspects of their daily living arrangements difficult to manage. It is not surprising to find that studies are beginning to recognize the impact of housing on people's health. Nor is it surprising to discover from our research that one of the great concerns people have in old age is that of where they will live and how they will cope with changes in their health or circumstances. For many, there is anxiety as to what the fit will be between their capacity and their housing; others, feeling freed from earlier constraints of work or family, are keen to develop new interests and live their lives in different ways.

Housing stories

We start the book with two older people telling their housing stories.

> **Mrs Crispin's story**
> I have a small room that I can escape to. In this room there is a large cupboard; it stretches from floor to ceiling and when I open it there are all my hobbies and craft equipment. ... Sometimes I can find solitude in my room and can sit peacefully when the windows are closed. It is not an environment people impose on me.

When I open the windows I can hear the noise of the do-it-yourself fanatics, of the children playing games in the back alley. There is lots of activity in the tree outside my window. ... I can smell the barbecue from the neighbour across the back ... In the winter I can smell the smoke from my neighbour's coal fire and it stirs my memory of times past. ... We need something to stir our memory as we grow older. ... Books are propped on the shelves; some are old and battered and smell musty. I love books, they remind me of a time when we had none, only comics. ... My room is like a sanctuary. There is a notice-board displaying old postcards and photographs of places visited in times past and happy memories when we climbed hills in the Highlands of Scotland and camped in the Lake District. ... Everything is higgledy-piggledy. In my room, I do not have to tidy up; I can make as much mess as I like. All the familiar objects surround me and help me to make sense of the world. ...

I do not choose to live in this house, I have to 'stay put' because I cannot afford to move. I have often pondered this situation and have grown quite attached to my small terraced house. In my younger days I always wished I could live in a dream home. But now I am older my outlook has changed. ...

When I look out of the back windows the scene it conjures up in my mind is Coronation Street. There is not even a blade of grass growing in our back yard but there are flower boxes surrounding the walls made from our old bed. There are enough rooms in the house for me to wander. ... There is enough room in the yard to do projects outside with the hammer, nails and paint. My touch is all around my home and I always like tackling new projects. My house is a riot of colour. My husband tells me he is getting quite used to that red standard lamp and orange shade with decoupage. Life is a lot more interesting.

I have put my mark on the house and identity. Home means security, a place to feel safe and comfortable. My life revolves around my home; it is a place to come to, a base. I can have family and friends to stay. The house contains all my personal belongings. ... I have privacy too.

The silver cup belonging to my father sits on a shelf in my room. Sometimes I pick up the tarnished cup and remember when it was clean and bright and shining in the display cabinet and of his life, so full of promise – the inscription reads 'amateur sportsman of the year'. ...

The powder bowl looks old, worn and faded as it sits on top of the bookcase. My mother worked as a waitress and looked ever so smart in her black and white uniform. She had jet-black hair and wore it in two pigtails wrapped around her head. Every day she would use the powder bowl and lipstick. ... As I watched my mother get ready for work each day as a child, I wished she didn't need to work and could stay home. So the powder bowl is a memory that stands out in my mind and still contains old powder I can smell and think of my mother. ...

I am now 92 years of age and living in a lovely bungalow with all the latest technology. ... The planners and architects listened to what us older people wanted in housing for later life. ... I no longer live in an area with people of different age groups and lifestyles. This complex is for older people. My old neighbourhood became run-down and there was not the same friendliness. ... We have a nice community spirit in our complex. We also have an entertainments officer. The bungalows surround a lovely garden. The houses are built so we can get outside on our wheelchairs. ...

I am now used to talking to buttons on the wall when I need help. When I want a cup of tea made I just press buttons and the machines come to life. Thankfully we also have care staff and this enables me to have my independence and live well. I still have some of my keepsakes around me but have a smaller cupboard. I also have an extra room for my family and friends to visit. ...

So there is more of a variety of choices for older people today. Housing is not being foisted on to people with little choice. Not everyone wants the same thing. Some people do not want to live in homes with rules and regulations; others are lonely and need to be cared for and like the company of others. There are plenty of services enabling people to live longer in their chosen home. (Mrs Crispin, 64 years, married, 2-bedroomed house, Lancashire)

In the second narrative, someone describes her thinking about where she would live as she got older:

Miss Membler's story

Whilst still at work as a resident college lecturer, I bought a bungalow which I regarded as a retirement home. After thirty years in the pleasant suburb I became convinced I must move and spent the next two years in real heartache as I tried for another home. The bungalow itself suited my needs perfectly. But when I became unable to garden myself, every summer was the same struggle to get gardening help. Added to that, the neighbourhood was hilly, and with age I

> found walking more difficult and I knew my time left as a driver was getting short. There was only one shop for food on the estate. Over that length of time the neighbourhood had radically altered in character. All these reasons made me look for a flat. So after looking at some flats advertised, I settled for a two-room flat in a [private sheltered housing provider] development, near enough to shops to walk there, on a bus route for the return journey uphill.
>
> The flat has one room fewer than the bungalow and I felt very cramped at first but I realize now the space I have is adequate for my needs as both rooms are fairly large (the bedroom is classed as a double) and it was odd to get rid of surplus. The flats have a guest room which can be booked ahead. I had the joy too as compensation of buying some new furniture to fit my new surroundings. My solicitor bought the flat for me. I had not asked his advice but I am sure that he would have advised me against the move if he had thought it unwise. On the financial side there was no anxiety as the bungalow sold for more than the flat cost, the management fees for the flat go up annually so caution is needed over spending. Up to the present, I have not felt at home, though I bought, not rented the flat and felt I have lost the freedom I had in my bungalow. Perhaps it is still early days and I may settle as time passes. If I become unable to look after myself, I shall have to leave the flat. Housing is never for ever! With the longer life span we now have, I was foolish to imagine buying a bungalow before I retired would finally solve my housing problem. (Miss Membler, 87-year old, single)

Mrs Crispin and Miss Membler illuminate different aspects of the experience of housing in old age. Mrs Crispin tells of her hopes and dreams; she links past, present and future; she writes about the way she creates her home and what it means to her. In the final section she imagines herself in 28 years time, when she is 92 years old and paints her vision of what housing ought to be like in a world where planners and architects have not only listened to older people but also acted on what they heard. By contrast, Miss Membler reviews the process of planning. She had made plans. The place where she lived remained suitable for her in terms of the building itself. However she could not manage to do the garden herself, nor could she easily find anyone to do it for her. The position of the house was not satisfactory as her capacities changed: she could walk less well, the area was hilly and she thought that before long she would be unable to drive. In addition local facilities in the form of shops were not good and the neighbourhood had changed.

So she moved to a flat. She describes the experience of getting rid of possessions and of learning to live with less space. The move had been planned to ensure there were buses for her return from the shops. It had also been costed and was within her financial means, though she remained anxious about the escalation of annual management costs. For reasons that are not fully developed in this extract, she describes not feeling at home and writes of a loss of freedom. Importantly also, she considers that she will have to move if unable to look after herself and concludes that houses are not forever.

Finding out the views of older people

This book draws its evidence from a research study *Housing Decisions in Old Age* which was a partnership in research between the Department of Applied Social Science at Lancaster University and Counsel and Care, a registered charity providing information and advice for older people, and campaigning for improvements in the quality of care. The work was funded for three years by the Community Fund and finished in January 2003.

The project focused on older people's housing pathways in later life. The emphasis was on the decision-making process and the consequences of these decisions. At the heart of the project was a commitment to involving older people, as both consultants on the work of the team, and as researchers, who conducted interviews with older people on their housing decisions. The project team has also involved organizations from the public, private and voluntary sectors concerned with the provision of housing, housing advice and support for older people, drawing on their knowledge and expertise through an advisory group that met regularly.

Central to our approach has been shaping the research to respond to older people's perspectives and attempting to understand the role of housing decisions within older people's lives. The research has been designed and redesigned to respond to the suggestions of older people. The research strategy has been influenced in unexpected ways by older people's involvement. The emphasis on the 'story-telling' approach to understanding housing has allowed people's individuality to emerge.

The background to the study is important because it explains how the views of older people were collected and shows how the methods critically contributed to the richness of the findings and theoretical insights discussed within this book. A number of methods were used to gain a wide view of older people's concerns:

Face-to-face interviews with older people in their own homes;
Postal survey;
Panels of older people advising on project plans and progress;
Written housing stories collected via letters in local papers; and
True for us events.

Face-to-face interviews

189 in-depth interviews were conducted by older people who had completed the Lancaster University certificated Social Research Methods course. This course was designed and delivered by project staff, initially in Lancaster, with 13 students completing the course. Subsequently the course was repeated in London with a further nine students successfully completing. We set out to provide students with the knowledge and skills to undertake social research interviewing, providing an important component to the project. 106 interviews (56 per cent) were conducted within North Lancashire/South Cumbria and 83 interviews (44 per cent) within inner and greater London.

Postal survey

We received completed questionnaires from 563 people, 34 per cent of a sample of 1646 older people supplied to us by the Office for National Statistics. In addition we have also sought and gained the views of a further group of 75 people resident in specialized housing and care settings for older people, having identified that they were underrepresented in the main sample.

Older people's panels

We set up panels in five locations:

Lancashire – panellists from a predominantly rural population.
Tyne and Wear – a mixed group of older activists, engaged in campaigning and tenants of extra care housing.
Leicester – a medium-sized city in which people from the Indian sub continent and East Africa have been settled for more than thirty years.
Lambeth – an inner city area, and an established group of black and ethnic minority elders.
Weston-Super-Mare – a seaside town to which people consider retiring, with half the group living in supported housing.

The panels met on six or more occasions over a two-year period; they were asked to comment on our plans, to share their own understanding of the issues we were exploring and to examine and validate the preliminary findings. Members were asked also to read and comment on the interview schedules and questionnaires.

Written housing stories

Working largely through the medium of letters to the editor in the local and regional press, we invited readers to submit their stories of how they planned their housing for later life. We asked them to tell us about the issues and obstacles that presented themselves, the successes achieved and advice they might wish to pass on to others. We received 125 stories in response.

True for us?

We set up five final meetings with panellists and interviewers to test out our findings. These meetings provided valuable additional material. First we asked whether the findings rang true. Second we examined which aspects of the study were the most important to emphasize to policy-makers. The third approach was to pursue topics on which we wanted more information. Thus we developed our understanding of the place of finance in decision-making. In addition we decided to look further at what people wanted to create in and from their homes. Various methods were used in these sessions: group discussions; writing lists and reflective writing. For the reflective writing, a creative writing specialist developed a framework to help people reflect on their current home. She asked people to write notes as they:

> Visualize taking a walk around their house until they come to a place they regard as special: think about the meaning of that special place or room; what does it means to you? think about sensual as well as visual aspects – sights, sounds, smells or temperature; what feelings are associated with the place? (McCraith, 2002)

Later she asked participants to list the ten things they would want to take to a new house.

Using various methods

We used a mixed research method design to explore 'housing decisions' from a range of different perspectives. We sought out people who differed in terms of their age, personal circumstances and living situations and gave people the space to communicate either publicly or privately, inviting them to write or speak to us about their experiences. The use of a variety of data collection methods allowed insights to be gleaned from different perspectives, such as personal, biographical, cultural. Some research methods are better than others at reaching certain groups of older people, for instance, people from minority ethnic communities, or people who are housebound, have sensory impairments, or the older 'old' age group. We hoped a mixed-method research design would help us harness the strengths and compensate for the weaknesses of individual methods.

In this study, 'older people' were defined as those aged 60 years and above, and included individuals who were at different stages of making housing decisions, including the decision not to move. First, those who had yet to move during retirement were asked to look forward to what might happen in the future. They were invited to discuss their current thoughts and plans, if any, for the management of their housing needs, should their circumstances change:

how suitable was their current house and location?
what options they perceived were available to them either to stay put or move?

Second, those who had indicated they were actively thinking about whether to move or stay, discussed the influences affecting their decision. In particular we asked them to think about the factors that they needed to weigh up and prioritize. Third, those who had already made a move during their retirement were asked to look back: how did their distance from the decision influence their perspective on what it was really like to make their housing decision in later life?

The two narratives of Mrs Crispin and Miss Membler came about from different approaches. Mrs Crispin's story was produced following a creative writing session at the end of the project.

Following the meeting, Mrs Crispin worked further with the facilitator on what she had written. Her writing illustrates an approach in which people develop insights from their own reflections. The second narrative, Miss Membler's, similarly, draws on personal reflection, though in this case it came from a biographical story written in response to a request in local papers.

Two further sources of information, interviews and questionnaires, are illustrated in the very different perspectives on housing decisions that follow. In the first, the process of the interview helps to create the searching that takes place in the reflection.

Mr Cranborne's story

Mr Cranborne: Well my wife and I came here in February 1999; we've been here two years.
Interviewer: And why did you come here?
Mr Cranborne: Well it's a longish story. I went on working full-time till I was 70 and my wife was a couple of years younger than me. ... And she began to develop ... Parkinson's. So in retirement my principal job is looking after her rather than doing some of the things I wanted to do. And it's kept quite well under control for some years, because obviously that is progressive, as you know. And it got to a point where she was beginning to need a lot more help, couldn't cope with the house and the garden and so we sold up in south Oxfordshire and came up here because I've got a son who's ... at the university. ... My wife felt that she wanted to get away from where she'd been for 27 years; she didn't want all her friends sort of breaking up.
...
Attached (to where my son lived) was a separate cottage which suited us quite well. So we bought that and moved in. That worked very well for three or four years; she was deteriorating and at 89 they began to realize that I couldn't look after her any longer. I had morning help, ... the county council people sent up, helped her to get up in the morning and what not but, once they had gone, I was in charge for the rest, getting her to bed at night as well. But it got beyond me in the end, so we got her into a nursing home. Then I said to (my son), 'Look it's quite pointless me living in the cottage here and her being in the nursing home. Far better for both of us to go into a nursing home'. I shall be 91 in July and I can't look after myself all that well at that age. All the cooking, all the washing and everything else and I'm quite happy to go into a nursing home, that is go in as a resident, she would be a nursing patient. And we looked around and ended up in this place and couldn't have done better really. It really is very well run despite staffing difficulties. And the atmosphere is very pleasant. So we had one of the two double rooms they have got in the place, the upper floor. Bit by bit she deteriorated and much more nursing care was needed and then ... eighteen months ago she suddenly ... developed epilepsy ... and we lost her in April last year. ... This is just the very thing I planned for because I knew very well that she was most likely to go first. ... So by then of course, I had my feet under the table, I'd been here well over a year and have stayed here ever since.
Interviewer: Yes, yes, so you really did think about it all step by step?
Mr Cranborne: Well yes my decision to come in when she came in to here, I knew this would probably be her final shot, and it seemed

> to make sense. (She was in the nursing home) and they invited me to join her there; they'd got one double room which was quite hopeless. Whereas this place has only been open for six years and is purpose built, so we had the double room, it really was a double room and it was very comfortable there in a way. ... (My son had said), 'Well, why not move nearer to (the town)?' ... I was then twelve miles out, awkward if you have no car and anything. ... And both (my son) and I said, 'Well this is the one we've got to come to'. So that was on the Saturday; on the Sunday we brought my wife over ... and she agreed that we've got to move and that was the thing to do and in we came.
>
> ... And (my wife) used to say, 'What are you going to do if I go first?' I said, 'I'll stay here, why should I move? I'm fed, all the laundry's done for me'. I can still look after myself with washing and shower three times a week. ... And I give them very little trouble unless I fall ill and then it's all laid on for them to take charge. ...
>
> There are times when time hangs heavy. ... I mean normally there are four or five of us fellows who have a table in the sitting room over the porch, the first floor sitting room, it's called the quiet room there is no television in there. They had a table for five of us, sometimes six, all men, and they serve us there instead of in the main dining room, which is half way down the building. Because they realize that we are reasonably *compos mentis* and find it a little bit trying sometimes to go into the dining room where curious things go on from time to time. ... I try and get out, when we had that dry sunny spell a while ago I tried to get out after lunch and go right round the building just for the sake of exercise and fresh air and sit outside and do a circuit beating about and then sit outside. There is a very nice terrace at the back. The floor above is at ground level at the back of the building so you can walk straight out onto the patio up there.

In marked contrast, a picture of people's perspectives of their own decision-making styles emerges from the responses to the questionnaire.

Themes in housing in later life

Many of the themes that we examine in the book have been captured in the stories and questionnaire data, as can be seen in Table 1.1

the impact of personality on: whether to plan or react to events; choice of home; and constructing a home and a life style;
the meaning of home;

Table 1.1 Decision-making preferred styles

Preferred decision-making approach	**'I wish to plan now for possible problems which may affect my ability to stay put'**		**'I prefer to stay in my home until it becomes impossible to stay'**		**'I would like to start thinking about my future housing needs, but I don't know where to start'**		**'I would prefer to react to any housing problems as they arise'**		**'There is no point planning for things that may never happen'**	
	N	%	*N*	%	*N*	%	*N*	%	*N*	%
Strongly agree	132	(23)	317	(56)	36	(6)	113	(20)	74	(13)
Agree	178	(32)	146	(26)	98	(17)	306	(54)	184	(33)
Neither	148	(26)	33	(6)	207	(37)	53	(9)	86	(15)
Disagree	43	(8)	38	(7)	131	(23)	39	(7)	138	(25)
Strongly disagree	14	(2.5)	9	(2)	47	(8)	12	(2)	43	(8)

Preferred decision-making approach	**'I do not want to think about my future'**		**'I don't have the energy to move house at my age'**		**'You can't plan for things that only *might* happen'**		**'I don't know enough about what the options are to make an informed decision'**		**'I would consider moving before things became difficult, to avoid having to move in a crisis'**	
	N	%	*N*	%	*N*	%	*N*	%	*N*	%
Strongly agree	32	(6)	52	(9)	48	(9)	48	(9)	124	(22)
Agree	98	(17)	108	(19)	176	(31)	170	(30)	280	(50)
Neither	89	(16)	79	(14)	77	(14)	125	(22)	67	(12)
Disagree	231	(41)	204	(36)	171	(30)	138	(25)	48	(9)
Strongly disagree	71	(13)	77	(14)	49	(9)	38	(7)	11	(2)

Note: Percentages total less than 100% because some people failed to answer.

the part housing plays in people's construction of self;
the link between housing and the management of daily living;
the impact of neighbourhood, and changes in the locality;
the changing health of the individual or their partner;
the anxieties about what lies ahead; and
the assumptions individuals or policy-makers have as to what to strive for in old age, such as maintaining independence or staying put.

One way to understand the impact of housing in later life is to search for key factors that are significant to what appear to be desirable housing outcomes:

the house itself: access to and within the house, including wheel-chair access and stairs; appropriateness of key facilities (bath/shower; wc; kitchen); size and type of rooms; ease of maintenance; running costs, including heating;

the environment: ease of access to shops, GP and facilities that matter to the person (place of worship, leisure); the quality of the neighbourhood (pleasant, safe); transport;

social network: access to and privacy from others; systems to call for help; nearness to relatives;

individual characteristics and resources: personality, lifestyle; state of health and capacity; finance.

From lists such as these it might seem possible to compile a model to produce the best fit answer: various factors would be fed in; the options would emerge. Such an analysis relies on an assumption that housing decisions can be viewed simply as the result of a rational, objective, decision-making process. However, no decision can be fully understood by breaking it down into components. People, whether consciously or not, have to manage their past as well as their hopes and fears for the future. Perceptions, beliefs, attitudes, feelings and thoughts all come into play.

Structure of the book

We move now to setting out the way we have constructed the book, illustrating the core themes with material from the research participants. The lengthy introductions to the chapters use housing stories both to open up the topics and to demonstrate the sort of material on which we draw.

Chapter 2: telling older people's housing stories

In this chapter we highlight two distinctive features of our approach that shaped how we pursued the topic and arrived at our findings.

First, the research strategy was driven by our desire to use methods that ensure the perspectives of older people were central to the research findings. This led to us developing a research certificate to educate and train older people in research skills to enable them to conduct the bulk of our in-depth interviews. We outline the various ways we developed to ensure older people had opportunities to be involved, to contribute their knowledge, to advise us on our project and plans, to direct aspects of the research process and to identify the key messages to be communicated from the research findings.

Second, although, as we have described in this chapter, we used a mixed-method research design, the vast majority of the data we draw upon in this book is qualitative. We wanted to encourage older people to be able tell their housing stories in their own ways. Qualitative material is the most helpful in trying to unravel people's experiences and look at the meaning they ascribe to events. We do draw on the quantitative survey results where they are relevant, but it is not our intention to report upon these findings here. All the methods are written about more fully elsewhere. (Clough *et al.*, 2004; Leamy and Clough, 2004; Leamy, 2005).

Chapter 3: housing decisions in later life

Here the focus is on the context in which people make their housing decisions. A part of that context is the way in which old age is perceived and constructed: how is ageing understood? What impact do such constructions have on the way elders and others review housing options? Some ideas and values become dominant, for example, views on the best living arrangements in later life, fears of living in residential homes or the primacy given to living in one's own home and maintaining independence. It is worth noting that there is little clarity as to what 'independence' means. The next extract captures different elements of decision-making: how far should you take advice from others?

Mrs Bennett's story

I have had Parkinson's for 34 years and also now have osteo-arthritis. I live alone in a small bungalow, which I own, where a few aids like grip rails, raised toilet seat, special water taps had been installed free by the local council for my use. The biggest mistake made to date was listening to other people telling me what I needed or didn't need. With hindsight, I now realize I should have made up my own mind, in fact, only listened to myself. I am the one who has these afflictions, I am the only one who knows what I can or cannot do. In December 1999 I was offered sheltered accommodation. A warden controlled flat. So putting my own place up for sale, I moved to the new flat. Being told that it was my home and I could do as I liked.

> No one told me the warden could barge in when she liked (which I put a stop to). You were expected to attend all the arranged functions, which I didn't. I stuck to my guns and only went to what I wanted to attend. One was made to feel like a spoilsport, but many of us were adamant. Before coming to this place, I was treated like a normal person, doing everything for myself. No one squeezed my hands and said, 'There, there dear, would you like to do a bit of knitting?'
>
> In the beginning, I was quite happy to see the red emergency pulleys hanging, but the warden just rings for one of your own relatives to come, not the doctor. The final straw came when the warden said, 'I will be very upset if you pull the cord just to ask how much salt to put in the potatoes or what vegetables to cook'. The consequence of all this, is that I have taken my bungalow off the market and am returning by July 31, 2000, whilst I still have a little confidence left.
>
> I said to the warden once, I expect many of the residents have hidden talents – 'No' was the reply; 'Nobody here can do anything'. Well, I have discovered they can. By the way, the ages range from 58 to 106 but only three are male.
>
> My decision to come here was totally the wrong one for me. I should have made my own judgement. Going back to live on my own will not be easy. I shall of course miss the company of the many friends I have made here. It has not all been doom and gloom. We do have some entertainment and outings. But I will be so happy to be back in my own little garden, watching the collared doves, wood pigeons and squirrels. They don't treat me as though I have a screw loose. ...
>
> Please tell the powers that be that we old uns are people like themselves who have the same emotions and feelings as everyone else. In fact, we are human beings. (*Written housing stories* – Mrs Bennett, 70 years, Bristol)

Of course this is the perception of one person. The extract is used to show the feelings of this person, not to examine the reality of what was said by a particular housing manager. The importance is that some older people think they are treated as if they are only fit for certain sorts of activities.

Someone else, reviewing the draft findings, wrote that she thought the most significant finding for her was 'the acknowledgement and realization that older people are individuals, who have very similar needs and wants to other groups in society'. She continued:

> However, I have spent several years in voluntary representation of older people attending groups where it is quite plain that 'the

professionals' have got quite clear ideas about what older people need and want – usually without any consultation.

She writes too about what is, in effect, her own theory of ageing:

> As I have become older, the material things in my home have ceased to have the same value for me. Photographs, mementoes of shared experiences are now more important. ('True for us?' – Mrs Richards, 70 years, widowed, 2-bedroomed bungalow, Newcastle)

Chapter 4: understanding housing decisions

In this chapter we consider both general theories of decision-making and the specific models that have been developed to explain and predict whether and when older people will move. We suggest some of the limitations, omissions and taken-for-granted assumptions of these models, arguing that, although rational decision-making clearly has a role to play in helping us understand how older people arrive at their housing decisions, the dominance of this approach has excluded other approaches from being adequately considered. Theories, as well as the experiences of older people in this study, illustrate that there are other dimensions. For instance, sometimes people are equally or more influenced by their 'emotional intuitions' or by prolonged reflections upon what is important to them within their lives.

The theoretical ideas and debates about decision-making outlined in Chapter 4 are important, but may seem quite abstract and distant from people's everyday lives. Therefore, these themes are illustrated by interview extracts which attempt to capture the differences that spring from people's decision-making styles and from their ways of thinking about housing. For instance, it is valuable to question people in relation to their housing decisions, first, as to whether they believe planning ahead is possible and, second, if they do plan, as to how they go about this: does their planning result in them waiting until circumstances change? Are they able to reflect on their reasoning? We begin this exploration here, with a quotation from Mrs Harrison, who in response to being asked about the planning of housing decisions, spoke powerfully of ideas that raise their head and go to sleep again:

> *Interviewer*: So, when it says how long have you been planning this move, it's sort of subconscious planning?
> *Mrs Harrison*: Yes, yes, and I think seriously it would be say 10 or 11 years when it sort of became a possibility, a serious possibility. And

> then add those years up to me retiring when I thought, 'Well I can't do anything while I'm working, I've got to keep working at the moment'. ... It sort of raised its head and then it went to sleep again and then suddenly I decided I'd had enough and I wanted to pack it in ... and think about what I was going to do. It was a bit scary actually because doing it on your own is bad enough if you're only moving two miles down the road. And I think if I'd been moving a long, long way to where I was not very familiar it may not necessarily have been the same.
>
> And I've since developed osteo-arthritis in my hip and I can see that being a problem later and people have already said to me, 'Oh, you're all right; get a stair lift fitted you'll be all right, not to worry, you don't need to move to a downstairs flat'. And I just think, 'Yeah, it's fine, it's in walking distance of town, it's within walking distance of everything'. (Mrs Harrison, 64 years, divorced, living in own house, Lancashire)

This person teased at the idea of moving for some time, like playing with a loose tooth, and then put it aside. But what in some ways is presented as a sudden decision is the consequence of the earlier incubation of ideas, which can be seen as a rehearsal. Another person, while not planning ahead, has in her mind the difficulty she will face if she cannot walk.

> .... That's one thing that worries me. It worries me an awful lot because I have found things getting a lot worse than it was in the first place. But I don't like to take drugs, I take an aspirin a day, but I hate to take drugs – I think I've kept pretty healthy. But, evidently in the left leg, from the groin to the toe, the artery was narrow and when I went to the specialist – I thought the veins would compensate for it, but I'm fit otherwise. The future worries me. I have two sons, one ... I don't see him much and the other one ... is very good to me. He comes on a Friday and takes me shopping and on Monday, takes me for a run and a walk. But, again he has his own family. But I have good friends, but I do most things on my own. That is the snag with being on a fixed income, I can't afford to have anyone to come and do anything for me. I have to do my own work. That is fine. On days that I don't feel well ... I don't do it. I sit and I read my book, do the crossword and things. I think, 'Right, the work will be there tomorrow and I can do it then'. With the lawns, I can do the lawns, then I have to sit down and I do everything in bits, but it's done. (Mrs Noakes, 87 years, widowed, residential home, Lancashire)

This is a picture not so much of planning but of musing over living arrangements, perhaps a combination of reflection, intuition and rational planning. Coming to a decision is a complex activity. Not surprisingly those living with partners are acutely aware of not knowing who will die first, or their respective health prognoses: if only this were known, people could better plan for each other and for themselves.

Inevitably, people living with partners are taking into account other people's feelings and wishes, not just their own, and, specifically, are aware that one of them will die first. Sometimes one of the partnership plan more for their partner if they were to die than the other partner may do for them. In mixed sex couples, sometimes this is men, who know that they have handled money and administration, or heavier tasks around the home; on other occasions it is women who are aware that their partners have done little cooking, and general housework, perhaps washing clothes and deciding what to wear. Some do have an idea of which partner may die first.

> But we decided that we could manage and my greatest thing was to make sure that, if I went first, my wife would be secure and as safe as we could possibly make it. (Mr Attwood, 79 years, married, 2-bedroomed flat, Lancashire)

For others, the consideration is more general: 'What will happen to either of us, when the other dies?' This is significant because of the capacity of couples to be mutually supportive in ways that may call into question the ability of either person to manage on her or his own.

> It may mean a progressive move to sheltered accommodation or even into care eventually, because sooner or later one of us is going to die – we don't know which one will go first – so what will happen to the other one? And it may well be that the one remaining will eventually finish up in care. This is one of those things that we don't really know what is going to happen. We try to foresee it and hopefully we can make some preparations beforehand for that time. (Mr Guy, 72 years, married, living in own house, Lancashire)

One couple mention various aspects of planning. First they say they do not want to look into the future:

> My sister and I were discussing it – she's a bit older than me and she's had a lot of ill health and she was saying, 'You would like to know

> what's going to happen, in view of what you need to do and how prepared you need to be'. But then you think to yourself, 'No, I don't, I don't want to know what's going to happen'. You don't do you? But then again ....

Indeed, they recognize the limits that there are to planning:

> Well, what we did ... we've a downstairs lavatory; there's a walk-in shower. We had a telephone line put in upstairs, and a television thing put in. We thought, 'Well we'll look for these things that we may need'. ... But I suppose there's only so much you can prepare for.

They note the frustration that the very things they had planned around may change:

> Often I go down on my own and I get the bus back. We had two an hour. I could cope with that, that was fine, because if I'd just missed one I could pop into Littlewoods or Marks or whatever, have a look round and then go back to the bus stop. They've changed it to one an hour now.

However, they are aware that they do not want to leave a move until, in their eyes, it is too late:

> You hear of such a lot of people getting to their 80s with a big house and they say, 'Oh I wish I had moved, I can't face it now'. And we thought, if we're going to do it, we have to do it while we're fit and able to because it is an ordeal moving. (Mrs Daniels, 70 years, married, 3-bedroomed bungalow, Lancashire)

Chapter 5: attachments to home

We move in Chapter 5 to explore the different ways in which people feel that they 'belong' in the places they live. In a sense, we look at what keeps people where they are. In our research, we have seen a number of ways in which people feel emotionally attached to their homes: from the security and control that private ownership provides, to a home's ability to house treasured personal objects, to how homes adapt to (and reflect) a person's life, values, and even bodies. We also show that home is valued as a place that supports the activities and hobbies which makes people 'who they are', and the social networks within which some feel 'roots' and a sense of social belonging. It is these emotional contexts of attachment to homes and places within which housing decisions take place. In order for those decisions to be understood, one must consider in what ways people have become attached to where they live.

In the creative writing exercise, Mr Blake wrote about identity as he looked at the place where he lived:

> Home, for me means companionship. I sit in the lounge with my wife, opposite the patio window, or in the summerhouse. We each have the same copy of a cryptic crossword, and share the solutions as we find them. Occasionally we look up to glance at the garden. In summer, the trees will be in leaf, the borders and planters full of bright flowers, and the wild birds are on the feeders or the lawn. In winter, the beautiful shapes of the bare birch trees are fully revealed. There are more birds than ever, and two grey squirrels perform acrobatics in the branches of the trees. The lounge is warm and comfortable. Two sleeping Labradors add to the feeling of content and well-being, and possibly a sense of achievement.
>
> The house reflects my identity, because we have lived in it for 36 years. It has sprouted extensions and adaptations to cope with increasing family, and its needs as it grew up. It is still security for my family, with room to accommodate another family in case of necessity. It is easy to run and comfortable.
>
> My personality has been influenced by external circumstances, and I am a survivor. I do feel that I depend greatly on relationships with my wife and children, and their families. We have been married for over forty years, and I dread being alone more than I dare admit to myself. (Mr Blake, 67 years, married, 5-bedroomed house, Lancashire)

At a later stage in the exercise, he was asked to list the ten essentials he would take with him if he had to move house:

> My wife; the dogs; my wits; a good dictionary; the computer; a radio; my sense of humour; a really comfortable chair and bed; a super efficient fresh air system; the addresses of my family.

Chapter 6: worried lives

Instead of dwelling on how people become emotionally attached to their homes, in Chapter 6, we shift our focus to the emotional stresses, worries and general feelings of vulnerability that are often prevalent during housing decisions in later life. We look at the worries and fears of individual older people and those of family members: crime, finances, death of loved ones and potentially of oneself, falling, losing independence or status on the one hand, and concerns about the safety and well-being of a parent on the other. Our aim is to show the different distressing emotional contexts in which older people often have to make housing decisions.

Understanding these worries, and the increasing vulnerability many feel, is essential to understanding the reasons for certain decisions, and provides a powerful insight into the ways the artificial distinctions between the rational and the emotional in the decision-making process break down when approached from a more phenomenological perspective.

Mrs Chaplin's story

Mrs Chaplin had lived in her house for 52 years. Indeed, she said, 'I think there were still rations when we moved in here'. The whole tone of the interview is about worries and loss. Her husband had died 19 months before:

> He'd been in a nursing home for five years, so I've been on my own for seven years. I mean, I got used to it, well, you never get used to it, being on your own, not completely, but you've got to put up with it.

She spoke of the problems of trying to look after him in unsuitable accommodation. He had had to go into hospital because 'he couldn't cope with the steps' after his stroke. She was told: 'There is no way you can manage to get him in and out of here'. She had no downstairs toilet. The house was not suitable for wheelchair access:

> I have a back side step, but it's only a thin path and with a wheelchair, it makes it awkward, there's potholes and whatnot up there with the cars, it makes it very awkward. So they said he has to go into a nursing home.

She commented that they had not had children: 'But it's hard, and we've no family and it's hard when there's only been the two of you, when you lose one'.

She was finding it very difficult to get the house sorted and to manage not only the costs of repairs and maintenance but also having to get the work completed satisfactorily.

But at the moment, I'm in a mess, as you can see. I'm looking in my cabinet and my kitchen's practically wrecked. ... Well I put my name down on the list four years ago for windows, that's all I asked for, new windows – they were absolutely done for. Now this window I paid for myself because they couldn't tell me when they could do it, I had to go on a waiting list you see – a big double bay window and it cost a small fortune. This year, they came to see me and what wanted doing was going to cost more money than the trust could afford, so I've had to take a mortgage out on this house. I had no mortgage, but I have now, £10,000. ... To help to pay for the repairs – new windows, new toilet and washbasin in bathroom, new roof, new damp course and all down the gable end, new ceiling in kitchen and gas central heating. Everything has gone smashing up to gas central heating. And I'm over the end with it.

I can't get them to come back and see it. ... I had four days with no heat at all. It was on the Friday, when the men left here, it went off Friday night and it never came on again. I rang them on Saturday morning and they said, 'You'll have to wait till Monday'. I said, 'That's against the law, you're not supposed to leave me, an elderly person in a house with no heat at all'. Well, he got very abusive and I've never heard from him since. ... (The housing trust) have paid for extra electric points put in and windows – this they got on a government grant. The firm that gets the grant from the government is in Newcastle and they sub-contract and then sub-contract. Talk about passing the buck. So I've been in touch with the men who put them in, and the firm in Newcastle who gave me the number of the firm who they put in charge and I'm waiting to hear from them.

I told them, when I rang them up, I explained the situation and I said, I do not want the same men back. That's why all the rubbish is on the front; it's been there over a week. I've still got an old radiator in my bedroom they wouldn't take out and I've still got the old water system in my bedroom that they wouldn't take out because the hosepipe wasn't long enough to drain it. Now what is the point of the government giving somebody money to make me, I'm saying me, to make me comfortable in my own home, when they wreck it. I'd four leaks in two days. ... I appreciate the free grant to put me central heating in, but it's not fair that they give it to a cowboy to do it, is it? ... I was sat here, a fortnight and water was pouring through that smoke alarm. I'm very hard at making friends.

She talked then about people she knew:

> The friends I have are very few and far between, but I've had them for years. Understand my meaning? I can't make new friends easily. My husband could, my husband, he made a lot of friends, but he also got me talking to strangers.
>
> ... So to move away from this area, I would be lost. I would be fastened in my house, I mean, I'm fastened in now, I don't go out very much at all and I suffer from asthma and what not, and I mean, I go shopping twice a week and that's it. (Mrs Chaplin, widowed, house, Lancashire)

Chapter 7: the looking glass self

A key factor in those decisions with which people are happiest seems to be the extent to which people both know themselves and know the aspects of their lives that are most important to their well-being. A capacity to reflect on one's own situation seems integral to this approach. We argue that knowing yourself also entails being aware and able to respond to your own and partner's strengths and limitations. Drawing upon previous research on coping strategies and styles, we discuss and illustrate various problem-focused and emotion-focused strategies that older people have adopted to cope with the stresses and tensions in deciding whether, when and where to move. Finally, in this chapter we examine the extent to which people felt a sense of control over their housing decisions as this has previously been identified as contributing to whether people feel comfortable with what they decided.

It is clear from the evidence that people do think through their present situation and think about what they should do in the future. However, it is equally clear, not only that some people adopt this type of pro-active approach to a greater extent than others, but also that for everybody there are powerful psychological aspects to housing decision-making: people want the place where they live to play a part in helping them to live, and to be, as they want.

Mrs Taggerty visualized her present and her future.

> **Mrs Taggerty's story**
>
> I feel happiest in the garden. It's something I have created and continue to improve – it is constantly evolving. It is somewhere to use my practical skills and an outlet for what little creativity I possess. The garden is what I would miss most if I had to leave.

I feel secure in my home. It is a warm, safe place to return to, where everything is familiar and arranged as I like it. I enjoy working in one room and being able to hear my partner working elsewhere in the house – with the cats loafing about occasionally checking up on one or the other of us. I love to sit by the front window and watch the world go by (well, the bit that goes by in our quiet street). I feel homesick if I am away for more than three or four nights.

My life and identity My home has replaced work – which formerly took all my time and energy – as the way I present myself to the world. My bold garden design style, the strong colours and plain layouts used in the internal decor, all represent me as I like to think I am – a straightforward, no nonsense, unsentimental person. I notice that my home is different from those of most of my friends and this gives me a real sense of being an individual.

When I am 92 I visualize myself as living somewhere smaller with little or no garden but with a stunning view from my main windows – if I can no longer get out to enjoy nature then I would like to be able to enjoy it from the security of my own 'nest'. Since my partner would by then be 101, I would not expect him to be with me. My son also, since he has the gene which causes Huntington's disease, will also probably have died and, since he is unlikely to risk having children, I will be rather alone and I think I will be lonely. However, I have always been a 'loner' and believe I will be able to cope as long as a have reasonable health and my imagination – I have always been able to lose myself in worlds of my own. (Mrs Taggerty, 60 years, married, living in own house, Lancashire)

Chapter 8: preferences in living arrangements

There are strongly held views about what should be provided in specialist housing for older people. One person stated what she wanted:

closeness to shops and doctors;
a lounge;
a laundrette;
a gas central heating (not low peak electricity);
a shower plus bath;
a airing cupboard;
a large cupboard to put coats plus vacuum, brushes and the like;
a fairly large kitchen.

She contrasts that with her own place:

> The flat I live in has none of these. We have low peak electric heaters but only a small heater on the wall in the bathroom which you only put on when you have a bath, so it is freezing in there; we have no airing cupboard at all. If we want a shower we have to pay to have it installed. Since I moved to this flat I have had to have two hip replacements, I also have angina. I have 15 steps to get to my flat which is very hard to climb. I have to dry my washing indoors because I can't carry my wet washing. (*Written housing stories* – Mrs Banks, Bristol)

Many people are frustrated and angry as they search for the best housing solution. Why, we were asked repeatedly, is it presumed that older people should be content with such small houses?

> **Mrs Patterson's housing story**
> We are 68 and 65, both with osteo-arthritis and spine degeneration. We gave a house up when we were in our fifties due to problem neighbours and moved into a (housing association) top floor flat and had 12 years of contentment, but realized health was going to be a problem with the stairs. My husband had severe depression for about seven years after being in the flat all day whilst I was at work. When I took early medical retirement we started to look at ground floor flats and bungalows, to buy and rent. Eventually we settled for this three bed semi-house and my reasons for doing so are as follows.
>
> We really wanted a two to three bed bungalow, but the size of rooms was ridiculous. Why do designers think retired or elderly only need a twelve to fourteen foot box room for a lounge? Bedrooms would not take our two three feet by six feet six inches divans; there were always doors or built in units in the way. Kitchens only have room for one person and no storage.
>
> Just because people get older doesn't mean they don't have visitors, for meals or to stay. Flats are about the same in private complexes. People need a private garden space.
>
> Council or housing association flats are more security minded, but too small. One person is expected to be grateful to be given a bed-sit to live in. Bed-sits are unhealthy and depressive, no escape from four walls. I have known people move into housing association flatlets

and become asthmatic through being forced to mix in the lounge (communal room) daily with heavy smokers.

Also, some of these complexes have an enthusiastic warden who arranges outings. This is fine, but some residents are embarrassed, as they cannot afford the outings and believe me, there is nothing worse than sheltered schemes for gossip.

Private complexes of flats and bungalows have very expensive service charges; the majority of these only need a mobile warden or contact housing officer, as they have a link up with control for emergencies and the residents do get cross at the poor and expensive maintenance works carried out.

Also, the worry is when a resident dies, until the dwelling is sold, the service charges have to be paid, by the family.

I know my husband has improved greatly since we have had a small garden and he has a garage, which is his working space, nicknamed 'the shed'.

We would dearly love to see the following built for the future:

- Bungalows with sensible room sizes.
- Semi- or detached.
- Walk in showers.
- Higher toilets.
- Quiet bathroom fans.
- Better noise insulation.
- Sensible kitchen units – with drawers and walk in cupboards, not below the work bench. Wall plugs easy to get at, not hidden away in cupboards. Eye level plugs in the rest of the rooms, decent size windows to let in plenty of light.
- Small private gardens easy to manage with a shed or communal area (people need taps for washing cars and buckets; these rooms become very handy).
- Gas heating, preferably combi-boiler.
- No resident wardens, but connected to control for emergencies.
- Flats.

Given the emphasis we have placed on the importance of recognizing the individuality of older people, it might seem that it is not possible to write much on older people's housing preferences. In fact common themes emerge, such as involvement of older people in design or demands for more space.

Chapter 9: policy and practice

Finally we consider the areas to which this study contributes: the process of decision-making, theories of ageing, research methods and housing policy. Our contention is that the housing decisions which older people judge to be the most successful come about because people have discovered the essence of what is most important to them. There are benefits from mentally rehearsing what is wanted as well as reviewing the practicalities of different types of housing.

Indeed, housing decisions seem best understood if they are taken as one aspect of 'the journey of life'. As discussed in Chapter 4, the term *housing career* has been used to capture this although some think that *housing history* better reflects the way housing is part of life. Whatever words are used it is important to avoid presenting people's housing stories as if they follow a straight line, all neatly planned. Deciding where to live is a part of deciding how to live and, indeed, of how ageing is seen.

The implications of involving older people more fully in research are discussed in terms both of the impact on the research and the potential for fuller participation for elders as citizens in their communities.

The last section sets out considerations for policy: the provision not only of information but of the sort of advice that allows people to think through what they want; the demands of older people for higher standards in housing both in terms of space and of facilities; and the worries people have as to how to manage to look after themselves and their homes.

Mrs Aspen's housing story

My flat has been on sale for nearly four years. I could sell it immediately if I could find somewhere, within my financial means, to move into. Age Concern, Help the Aged, the benevolent fund of my late husband's profession have been approached but seem powerless to help.

All the housing associations have properties way above my means and in any case, a list of properties scattered around the county, when I have no transport, is pretty useless. I want to move into something nearer the town centre because although my present flat is only a mile away the bus service is threatened and the service finishes at 6 pm anyway. As I have no family to advise me, or talk to, I would welcome a service that could advise me on the cost of a move and on disposing of some contents of my flat.

> Four estate agents have offered to sell it at £1000 plus VAT but none of them has offered to help finding a place. In fact, they seem to do very little for their fee. The reason for my difficulties is, I think, that since the election, salaries have risen to meet high housing costs when pensions have practically stood still. Each year that passes I get relatively poorer. I don't know if any of this is what you are seeking in your survey, but it is very much a worrying matter for this 73-year old. (*Written housing stories*, Mrs Aspen, 73 years, Birmingham)

The 1017 elders who participated in this research want their voices to be heard.

2
Telling Older People's Housing Stories

Introduction

In the previous chapter we provided an overview of the research design. We now wish to explore two distinctive features of our approach that enabled us to gain different insights into the topic: first, the involvement of older people within the research process in order to make the research more responsive to the perspectives of older people and, secondly, the use of qualitative techniques to collect older people's housing stories. The need for alternative approaches to gerontological research has been acknowledged recently by other researchers:

> We want to ask for a different style of research than that often carried out. The key difficulty is that the vast majority of research into old age is conducted by 'not old' researchers. So often assumptions are made which are not always valid. Although a wide variety of methodology should be employed, the current bias towards measurement and structured approaches could be tilted slightly in the favour of a range of qualitative techniques which involve listening to old people's accounts. (Heywood *et al.*, 2002, p. 163)

Jerrome (1992) suggested that youthful researchers often avoid employing in-depth interviewing techniques because they find it awkward to meet old age head on (cited in Heywood *et al.*, 2002, p. 26).

Participatory research with older people

The older researchers, with ages ranging from the early 60s to 75, brought a considerable breadth of life experience to the research.[1] There were 13 women and nine men, and three were from minority ethnic communities. Whilst a few had left school when they were 14-years old, others had obtained professional qualifications or entered higher education and completed undergraduate or postgraduate degrees. In their personal lives, many of the group either had faced or were facing difficult decisions regarding their own future housing needs. Some knew of the problems vicariously, having looked after and witnessed the problems their parents had encountered.

From comments made on their application forms, not only did they welcome the opportunity to get involved in doing research, but also they saw how the research would offer the chance to pursue a topic of direct interest and relevance. For instance, three people stated:

> I have natural curiosity. I have developed an interest in this particular topic as a result of my work in the field – for the census. My geographical area took in two lots of sheltered housing – all home comforts, but sterile. Why?
>
> I enjoy working with people and their problems. Housing choices – making them instead of drifting – appeal particularly. Along with a number of people I know, I am facing the dilemmas myself and learning to approach and tackle housing decisions and helping others to do so could be of enormous benefit to the community and to me.
>
> I would like to do this course because I have an interest in older people, especially in their welfare and I feel that sometimes their needs are not always met or even considered. As a member of a team, researching on old people's housing, it would give me a chance to put their needs and requests forward.

The voice of older people themselves has largely been missing within the debate about older people's current and future housing needs and aspirations, although there have been some notable exceptions, such as the *Better Government for Older People* initiative. The *Housing Decisions in Old Age* research was a vehicle for older people to communicate their views on current housing provision to policy-makers, housing and social care professionals, architects and property developers. We believe that including older people in research and policy that will affect their lives is vital. Older people are often excluded, marginalized or subject to discriminatory behaviour within society and involvement in the social research process is an effective way of giving more volume to their voice, using research to help them to understand the world in order to change it. Underpinning the research strategy was a desire to relate to older people in an original way, by treating them as research colleagues rather than as advisors or research subjects.

Opening up the research process and inviting people who typically are the subjects of research and practice (often described as service users and consumers) to get involved has been variously described as conducting 'inclusive', 'participatory', 'empowering', 'emancipatory' and 'collaborative' research. The terms are used apparently inter-changeably, although the intentions behind the invitations to participate may be very different. Critically, the way the impact of involving lay people in the research will affect the process, the findings and the participants, will depend upon the original reasons for the invitation, as well the nature of the involvement that is on offer. If, as in empowering, emancipatory or participatory research, the explicit research aim is to empower the participants, then this is the driving force behind the design of the research, rather than empowerment being a welcome, but indirect, consequence of involvement. Such a research approach may include sharing decisions about aims, methods and conclusions, though De Koning and Martin (1996) warn against adopting a purist attitude. The intention of this type of research is to share power and work towards equality between participants who take on research roles and academic researchers.

There are similarities to some styles of community development in which activists gather local knowledge, in effect conducting a type of research, to bring about change in policy or practice. Information of what is happening is used to promote change. Indeed, Tasker (1978) describes community work as 'compensatory education'. There is another discourse on the role of community work in social planning, as distinct from simply involving people, which matches the perceptions of some interviewers, interviewees and panellists in the *Housing Decisions in Old Age* research.

Participative research fits within the interpretative research paradigm of qualitative, ethnographic and feminist research methodologies. In this, subjectivity is valued. Although the positivist research methodologies carried out within the tradition of scientific research can be participative, there is less scope to influence the process of knowledge production because the researcher's aim is to remain as detached as possible. There are models of involvement, which outline a progression of involvement stages, from deliberately or effectively being excluded from the decision-making process to being given considerable scope to influence and control decisions. (See, e.g. Arnstein, 1969; Taylor *et al.*, 1992.) Throughout the life of the *Housing Decisions in Old Age* study, the ultimate control and responsibility for research decisions always rested with the research grant holders, a structure which created many tensions. The scope for delegating this responsibility to older people differed depending upon the research method, some of which were better suited to providing opportunities for involvement than others, as illustrated by Table 2.1.

It is worth emphasizing that whilst the level of involvement may vary, we do not intend to give the impression that a high level of involvement is always preferable to a lower level. The inherent danger in using these types of hierarchy is that it may be implied that participatory research should always aim for maximum participation, and that, if it falls short of these ideals, it is judged as inferior. Involvement has to be a matter of individual choice. We know of panellists who recognized that their contribution to the whole project was relatively limited but, having heard about the extensive number of research interviews conducted by the older researchers, were nonetheless delighted and fulfilled by their participation. Rightly, they thought they had made an important and valued contribution. We had given them the opportunity to get more involved. For example, the London and Lancaster panellists were invited to apply for the research methods courses. Several took up the offer, but the majority were content to limit their participation to being panellists.

The research staff who facilitated the initial panel meetings would have been pleased for members to take responsibility for chairing or planning the meetings. Members did not want to do this, though they played a part in identifying the topics for discussion. A significant part of the reason why members did not take on greater responsibilities was likely to have been that that the panels were created for the research and therefore were not existing groups with structures, though of course some members might know others. The extent of involvement also fluctuated in response to the specific preferences of older people, some of whom chose not to participate in certain activities such as the literature review, the data analysis or the dissemination events. Some saw the tasks as unappealing, others said they felt they were not best placed to undertake these tasks and preferred to channel their energies into areas where they felt their strengths lay and could make most impact. In the older people's advisory panels, some people, most notably those who were already in some way engaged in campaigning, were initially sceptical about being involved and voiced concerns that no one would listen to their views and that research was not an effective method of delivering results quickly. It is important to note that during the three years of the

Table 2.1 Roles and authorities of older people within *Housing Decisions in Old Age* research process

Level of involvement	Role in research	Decision-making authority in research process
High	Novice researchers conducting in-depth interviews	*Had authority* and freedom to decide: • style of interviewing; • nature of relationship with interviewees during fieldwork; • choice of how to pursue topic and follow-up responses made by interviewee during interview.
		Had the authority to take *some* decisions concerning: • sample selection; • design of interview schedule; • data analysis of their own interviews.
		Had the authority to take *some* decisions concerning: • how to write-up and disseminate their involvement in the research process through their own thinking and presentation at seminars and conferences.
	Panels of older people	*Advising* on project plans and progress.
		Contributors of knowledge, but with some limits to what and how much could be revealed as a consequence of the group context.
	'True for us?' end events	*Contributors of knowledge*, but with some limits to what and how much could be revealed as a consequence of the group context.
		Had the authority to take *some* decisions, namely choice of main messages to highlight from research.
	Written housing stories/ interviewees	*Contributors of knowledge* with control over what and how much to reveal.
		No scope to contribute to wider research process.
Low	Postal survey respondents	*No influence over research process* as in role of traditional research subject. Contribution limited and determined by nature of research questions, though there was space at the end for them to say anything else they wanted.

Source: Adapted from Taylor *et al.* (1992) 'Levels of Empowerment'.

research, there were significant health problems for some members or their partners.

We now move on to describe one of the specific ways we developed of working with older people.

Co-learning: an example of sharing skills, knowledge and experience

> It's not just a course with you teaching research methods, but everybody learning together. (London research student)

As part of the *Housing Decisions in Old Age* research, a 'Certificate in Social Research Methods' was designed by Lancaster University, restricted to people aged 60 and above. Developing a course in research skills was one component of the attempt to relate to older people as research colleagues. The students came to be regarded as team members, though it would be inaccurate to suggest they were full or equal partners, as they did not have formal responsibility for the research outcomes. Nonetheless, they did share some of the pressures and anxieties associated with doing research and were thoroughly committed to ensuring the high quality of the research. For their part, students contributed their previous experience, knowledge of interviewing and of old age. They constantly challenged research staff to question assumptions, processes and methods.

The course combined theory and practice, in an attempt to share academic research knowledge and practical interviewing expertise in an accessible way. The first task was to demystify the research process. Emphasis was given to the importance of ensuring that the choice of research approach, whether that be quantitative, qualitative or a combination of both, is determined by the nature of the research question. The certificate was pitched at the level of an undergraduate degree course and enabled students to develop practical skills and theoretical knowledge, as well as to increase their self-confidence. Unlike other continuing education courses, where students only have contact with tutors during their allotted teaching time, students were encouraged to contact tutors at any time if they were facing difficulties related to their fieldwork.

Students were required to pay course fees, though these were heavily discounted as the course was subsidized from research funds. In addition they were paid for conducting fieldwork interviews that formed part of the course requirements, a practice that has become more common for consumers attending research consultation groups, but is rare for students of research methods. This meant that they were able to recoup the cost of the fees and, indeed, be in profit at the end of the course.

Given the theoretical backgrounds of the two course tutors, the course was designed to teach and promote the use of feminist research methods. The general aims were to raise awareness of the scope of quantitative and qualitative research methods and help students develop an appreciation of real-life social research considerations. Specifically, staff aimed to teach practical skills, such as interviewing, qualitative data analysis, ethical practice and customary research etiquette. Embedded within the explicit course content was an implicit desire to challenge previously held assumptions about the nature of academic research. Initially, many students could not conceive of research which did not follow the scientific method and thought research should be carried out only by distant, neutral and objective scientists. Consequently, they thought that it had little relevance to people's everyday lives. To introduce students to alternative ways of doing research, the tutors encouraged students to read and discuss articles on the merits and drawbacks of peer interviewing[2] and qualitative research.[3] These articles allowed topics to be examined in a practical way: how and when to use self-disclosure; how to invite interviewees to take more control in the interview; and how to undertake research that is ethically sound. The written assignments specifically encouraged students to develop their ability to reflect upon their own development and research practice.

When the programme was completed in Lancaster, it was repeated in London, with minor adjustments in the light of experience. The task for research staff became even more complex. It involved delivering a course in a different format (different length sessions and mixed mode teaching – face-to-face teaching and distance learning) and supporting two different cohorts of students at different parts of the process. At the same time, staff were also contributing to other research activities.

A further unexpected dimension emerged when the first cohort of students, having completed the course, conducted the interviews and received the award, told us that they wanted to continue doing research. Their appetites had been whetted, they felt somewhat abandoned as the staff focused on other parts of the research and they asked for further training. The research team tried to support them in making use of their new skills, energy and enthusiasm, and developed further training courses. So, the project, which had not set out to become a job creation scheme, had an unplanned consequence.

At the time of writing ten of the students in the North West have set themselves up as a co-operative consortium called *Older People Researching Social Issues* (OPRSI) and are now offering their services as qualified, experienced social interviewers. It seems likely that they will be seen as an excellent resource for public, private and voluntary sector service providers to draw upon in eliciting consumer views. Increasingly, the perspective of older people is seen as central in both evaluation and development of policy. Since the end of the *Housing Decisions in Old Age* research, OPRSI have now lined up research commitments for the next three years. They have just completed a one-year project, funded by the Joseph Rowntree Foundation called *Older People as Researchers: potential, practicalities and pitfalls,* and are to join in a two-year project on *User-professional relationships in social and health care'* managed by Glasgow University and funded by Department of Health. They have also written research proposals, undertaken smaller research contracts with local charities and authorities and participated in training activities.

Some lessons learnt on involving older people in research

There are many implications to involving older people in research, some obvious, others less so. We agree with Peace who has argued that 'Research involving older people needs further unpacking' (Peace, 1999). Whilst there is not space to do justice to the topic within this chapter, we shall highlight some of the pertinent issues.

We believe the distinctiveness of our approach lies in the integration and co-dependence of research and teaching practice. We know of a few recent or ongoing examples of running courses to equip older people to carry out research, but they require participants to initiate their own research projects.

By merging research and teaching in an unusual way we were able to give students the opportunity to be engaged in a real life, ongoing social research project that served to unite them in a common venture that had credibility, a clear set of aims, objectives, timetable and outcomes. The research topic, whilst not originally selected by them, was recognized as being pertinent to their own lives, and clearly motivated them to complete the course and the research tasks to the best of their ability. Their commitment and engagement with the course, also arose from being students and wanting to learn and perform as well as they could. To date, their four-year journey has taken them from lay people, to students, to novice researchers and on to being more experienced researchers.

From our perspective, the scale of the research activity, with 200 interviews conducted by older researchers in Lancaster and London, was ambitious, but we hoped it was evident to students and outsiders, that we were taking their participation seriously. It was a great leap of faith with a lot at stake for students and staff, though the anxieties of students and staff differed: students took the same risk as any student of not passing a course; staff knew that high-quality interviews were vital to the success of the project, with reputations as academics and competent researchers at stake. Student dropout and failure took on whole new meanings and staff had to face the question of whether this approach would really work.

The setting of the course within a university was significant. Learning in such an environment was a way of creating a new identity and status of 'university student', as this comment reveals:

> The *Certificate in Social Research Methods for Older People* had all the elements I needed to help me adapt to a new and interesting lifestyle, as I grew older. My street credibility with my younger friends is something to die for! I have discovered a new language, a new and complex subject, rediscovered my intellectual confidence and re-established my feelings of self worth. I was very much a beginner, but the effort needed to produce work for the course was a steep learning curve and my skills improved daily and, with them, my choices.

The course was validated by Lancaster University and worth 40 credits towards an undergraduate degree. As it was a very challenging course that demanded considerable commitment and high standards, students rightly experienced a sense of accomplishment. The course aims were to *educate* students in research theory, methodology and techniques, rather than simply to *train* research participants on a short course on research skills.

We learnt many lessons about the process of involvement and lived with many tensions. Here we highlight a few key attributes and dilemmas.

Teaching dilemmas

Teaching skills The research staff needed to know how to teach effectively to enable them to share their research skills and knowledge. One of the staff took the university's Certificate in Learning and Teaching for Higher Education.

Reading transcripts of student interviews For academic researchers, this was a twofold task. Each transcript was read to give the students critical feedback on their interview style and techniques, and a verdict on whether their interviews had been useful. From a research perspective, the reading was aimed at becoming familiar with each interviewee's story, then analysing, theorizing and developing ideas from what had been said.

Team meetings Part of the teaching sessions was given over to discussions about the research activity and the students' part within it. We wondered whether holding team meetings of research students and academic researchers, separate from teaching sessions, to share ideas, solutions and thinking might have been preferable. By allowing the discussions to take place during teaching sessions, boundaries were blurred between roles as teachers and researchers, which had advantages and disadvantages.

Supportive supervision sessions It was vital for students to have access to peer support and research supervision. Group sessions were scheduled into the research timetable to deal with fieldwork issues as they arose. For instance, the interviewers had a profound sense of responsibility towards interviewees, with some being drawn into situations where they felt over-extended, helpless, emotionally exhausted and unsure how to extract themselves from their involvement.

Language Church (1995) identified a mismatch in communication styles between professionals and users which needs to be overcome for their collaboration to become more effective. Rather than adopt the position that academic language should be banned entirely, we tried to minimize it and, where its use was unavoidable, explain each term at the time, handing out glossaries of research terminology as an *aide memoire*. Some academic peers questioned whether we were in effect, 'turning them into us', replacing their amateur, lay status with a new position as beginning research professionals.

Theoretical issues

Within qualitative research there are ongoing debates about representation (how to capture and communicate lived experience), legitimatization (how to evaluate qualitative research) and praxis (how to change practices and policies in society). Whilst there is not space fully to develop these theoretical considerations here, we have picked just one, representation, to illustrate some of the dilemmas and issues.

Academic researchers are 'doubly distanced' from interviewees Where academic researchers do not have direct contact with the interviewees, they are effectively doubly distanced from the research participants. Their world is viewed through the eyes of both the older interviewers *and* the academic researcher, compounding the difficulties of attempting accurately to interpret and represent their words. The 'True for us?' events provided some interviewees and panellists with the opportunity to say how their stories were represented within the research report. Unfortunately, it was not practically possible to extend this opportunity to all research participants. There is a tension for academic researchers between wanting to have direct contact with the interviewees to develop their own knowledge of the research topic and wanting to work with older researchers to develop their skills in research.

Insider researchers Researchers who have inside, personal knowledge of the research topic have a 'double knowledge' and 'connections by birth or from an earlier life experience' of the culture in which the fieldwork is being conducted (Okely, 1996, p. 25). Similarly, Callaway refers to this as having a 'double frame of reflexivity' (Callaway, 1992, pp. 32–3). Researchers never enter the field as neutral or impartial observers, but arrive with extensive baggage and immediately negotiate complex issues about their positions. The older researchers came to the research with multiple roles, not only as older people, but also of living on their own or with partners, or of being parents, grandparents or without children. Further, some had been born in the UK, while others had been born elsewhere. Of course their state of health varied markedly. They had differing pre-existing ideas and beliefs about whether people should move in later life, and, if so, under what circumstances, and differing views also on what they would do, or had done, in various circumstances.

Practical and organizational issues

Management demands The management and organizational demands placed upon academic researchers were substantial, in that they had to oversee the wider project and the in-depth interview analysis, as well as supervise the teams of interviewers.

Information and advice The interviewers needed to understand enough about housing in old age to be intelligent listeners and questioners. Research staff and students were concerned about what help would be available if and when interviewees described some of the difficulties they were facing. So an information sheet was produced which, amongst other details, gave an address and phone number for *Counsel and Care* in case individuals wanted further advice. As all researchers, the interviewers were working out with support from research staff, ways of appropriately managing the roles of researcher and concerned citizen.

Time The time taken to teach research methods and supervise fieldwork could have been used for academic researchers to do the tasks themselves. There is a need to build in sufficient time for students to learn by doing and making mistakes.

Keeping everyone involved The process of translating the ideas and plans from the outline in the original research proposal into practice took considerable thought and frequent discussions amongst the whole research team. An overall research strategy and detailed methodology evolved slowly and was constantly amended to reflect changes in circumstances and unanticipated events. Naturally, during the lifetime of a research study, the research team developed and refined their understanding and knowledge of the chosen topic and this affected the approach to the work. Opening up this process to include people who typically are the subjects of research was challenging for everyone, but we believe the rewards justified the effort.

We now move on to the second distinctive feature of our approach, the use of qualitative research methods to allow older people to tell their housing stories.

Telling stories

The majority of the data that we have drawn upon in this book is narrative. By deliberately choosing to emphasize the narrative details of older people's everyday lives and experiences, the research aimed to inform policy-makers about how older people approach the task of making housing decisions, how they construct their lives in their homes and what they look for in housing. Elders were encouraged and given the freedom to tell their housing stories in their own ways. Research methods and approaches were chosen that would allow diversity of experiences and choices to emerge, rather than disguising such complexity by attempting to reach a consensus on how and why people make the housing choices they do.

The narrative, or interpretative turn, places importance upon meaning, context, time, reasonable accounts, and the normality of inconsistency and emotionality (Gubrium, 1993). Sarbin (1986) views *story* or *narrative* as a way of organizing episodes, actions, and accounts of actions in time and space. He believes that narratives help organize fantasies and daydreams, unvoiced stories and plans, memories, even loving and hating. Using such a methodology to concentrate upon the subjective meaning of people's decision-making experience, and ensuring it is placed within the wider narrative context of people's lives, allows very different stories to emerge. The focal point of the story is not restricted to the present, or to housing needs, but can be enlarged to take a wider view of the meaning of housing decisions in relation to life as a whole.

Stories are socially constructed and shaped by the questions that are asked, the people who ask them, the way in which they are asked and the purpose of the question. For research interviewers, the consideration uppermost in the mind is whether the interviewees' stories shed light on the research questions they are trying to answer. Whilst a story is being told, research interviewers must be continuously alert to the need to prompt for elaboration, clarification and explanation. There must be checks to ensure they have understood correctly and appreciated the significance of what they have heard. There is often an underestimation of the interviewing skill required to decide whether to pursue a topic or move on to another question. It is not always apparent if and how a tale is relevant to the research question, and patience is often rewarded when the links are finally revealed either during an interview or when reading the transcript. Crucially, from the interviewee's perspective, the motivations are different. Self-narratives enable a sense of self, other and place to be constructed and reconstructed. Using the metaphor of the person as a motivated storyteller, Hermans (1999) suggests people, in telling stories, strive for self-enhancement and self-investigation. He notes that people 'do not tell their stories as though they are exploring a free space, but focus on those parts that arouse affect or even strong emotions' (p. 1193).

For some people, the request to tell their story may come at a time when they are preparing answers to their own pertinent questions. Asking people for their predictions about what they *may* do in the future is problematic. We have long known that behavioural intentions and attitudes are not reliable indicators of future behaviour (LaPiere, 1934), but such questioning does reveal how and when people approach the task of making housing decisions. For instance, they may have been thinking about the issues, but may not yet be able to answer conclusively. So they may give hesitant, partially formed answers, triggered by their own leading questions and by conversations with friends, observations of others in similar situations, or influenced by social norms or stories in the media. This should not be a surprise. Gubrium (1993, p. 6) reflected that 'as researchers, we diversely construct, deliberate over, debate, and periodically reformulate our own ideas about the lives we study. Why should we suppose that those studied don't do likewise?'

Whilst some people often give cautious, provisional explanations, qualifying their replies with 'it depends', others may simply not have been thinking about the topic at all, only considering it when prompted by the interviewer. Their stories show how often they change their minds, experience conflicting emotions and have concerns about the ways others, including family and friends, may view them. Sometimes, to an outsider (which may include family members), their intentions, attitudes and behaviour only make sense when viewed within the context of their lives and experiences. What is apparent is that people's housing stories contain contradictions, inconsistencies and ambiguities. These are not necessarily unreasonable. When people give 'it depends' type answers, their replies have to be placed within their 'narrative context', paying particular attention to how the context affects feelings (Gubrium and Lynott, 1983). Crucially, 'it depends' type answers cannot be captured accurately using fixed choice questions.

We now move on to trying to convey a sense of how we developed an ongoing understanding of the topic, together with the kinds of questions, issues and dilemmas that arose.

Overview: our understanding, analysis and interpretation

Developing our understanding, analysis and interpretation of the data was an ongoing process. Throughout the lifetime of the research, we shared our reflections at research team meetings, supervision sessions with the older researchers, the older people's panel meetings, and the series of one-off end of project 'True for us?' events with panellists, older researchers and interviewees. As we became more familiar with the topic, we realized that our overarching theme was individual *housing pathways*, acknowledging that each person might make more than one housing decision in later life. For some, their housing pathways might be represented by a steady line, but for others the process might be better seen as a jagged line, with crisis points requiring different housing decisions. We found we also needed to be clearer about what was meant by a 'decision'. We dispensed with the notion of a decision as an absolute event, a moment of choice, shown for example by a statement, 'We go here rather than there.' We came to see a decision as an accumulation of repeated reflections, perhaps triggered by life events or conversations.

We wished to explore how people contemplate their future housing decisions, with what we called 'What if?' scenarios.

- Did they discuss the decisions with others and, if so, in what way?
- What nudged them into this style of thinking?
- What happened when people nearly moved, but then decided not to?
- How did they interpret their situations and what choices did they make in response to them?

In particular, we were curious about the ways in which people responded to a range of different circumstances, events, hardships, losses, dilemmas and conflicts of interest. We wondered how they balanced and assessed the various underlying risks of each course of action. How did the reasons for staying or moving home accumulate and interact?

For example, declaring that 'We would like to stay if we can cope' may be fine in the context of current house and current environment. In the future, however, if someone cannot drive, poor local transport may make staying put impossible, so it is not just that being unable to drive leads to a move. Being unable to drive becomes significant in a particular place because transport is so poor. So we searched to find what people said about the opportunities and options that they identified as being available to them.

What did they find difficult to discuss when recounting how they reached their decisions?
What strategies, tactics, coping skills and resources did they use?
What solutions did they find to deal with their circumstances and situations?

Early on, we found that many people objected to the use of the word 'planning' because it misrepresented what they were actually doing, implying as it does a cool, objective, rational, systematic process that results in definite sets of plans for different eventualities. However, they would describe reflecting and mulling over possibilities, imagining and rehearsing potential future scenarios: sometimes wishful thinking and daydreaming, but at other times being more realistic. One person noted the word 'plan' itself was problematic and helped perpetuate confusion.

> The use of the word plan is the trouble, because the word plan can be used and is used in this statement as something that's desirable and something which exists. A plan implies that this plan is in being,

> but in parts it isn't. A plan is something that you hope for, but plans themselves can be much more optimistic than is the fact.

We conceived the process of decision-making as encompassing several thinking processes: rational, intuitive and reflective. Housing decisions are an intricate web of people's hopes, expectations, aspirations, dreams and beliefs about the future. Choices are underpinned by emotional and biographical reasons for staying put or moving: retaining one's dignity, respect for oneself, pride, ability to sustain and make new relationships, potential for retaining or developing new lifestyles. We tried to identify the distinctive aspects of older people's housing decisions but, just as for other age groups, selecting housing in later life is also about choosing a lifestyle, not just judging whether the house and its location can provide ageing-related needs.

We looked at individual accounts to compare and contrast ways of dealing with common experiences to see what could be elicited, but were careful not to over-generalize. With many social science research methods, there is a tendency to reduce individual experiences in order to repackage them as essential variables to be presented numerically as averages, percentages, predictions and probabilities. The danger with such approaches is that the coherence of everyday lives is lost. By comparison, qualitative analysis is better suited to preserving the connection between events, sequences and actions, as well as the emphasis the storytellers place on aspects of their story. Yet within qualitative research, there are still analytic tensions. For example, on the one side, there is the desire to retain the flow of each story by reporting it in its entirety but on the other, a need to adopt a more systematic and analytical approach to describe, conceptually order and theorize (see Strauss and Corbin, 1998). Einagel notes the tension:

> I have found it impossible to subject my respondents' narratives to forms of analysis that cut across the thread of biographical story telling. The continuing imperative is to preserve biographical integrity ... drawn explicitly from only one person's story; ... to quote from several would feel for me like a violation of their selves. (Einagel, 2002, p. 234)

Within this book we have opted to used thematic analysis to analyse systematically the meanings people were communicating, looking at both the manifest content and also the latent content to discover what was implied. Therefore we were able to take a cross-sectional view of specific topics across all stories; but we have also tried to stay with people's storylines as a whole by giving a fuller account of their individual situation. In doing so, we recognize the dilemmas that Einagel discusses but accept that compromise is necessary.

Conclusion

We have tried in this chapter to provide an insight into the way the research strategy has crucially shaped our thinking and findings. For instance, we found giving people space to tell their stories in their own way did indeed motivate them to 'focus upon those parts that arouse affect or even strong emotions' (Hermans, 1999, p. 1193). As a direct consequence, the emotional context of housing decisions has emerged as a key over-arching theme to our findings.

Although we did not seek explicitly to collect biographical data and long housing histories, for some people the data does allow us to trace links between previous life events and experiences that directly and indirectly shaped subsequent housing decisions. We did not tell interviewees where to start, but allowed them to decide. This turned out to be a considerable bonus. Some of the older researchers did explicitly seek their interviewee's life histories, partly out of a natural interest and curiosity about other people, but also because we had encouraged them to feel free to develop the interviews in ways that seemed appropriate to them. At first, we advised against dwelling upon interviewee's life histories because it seemed that the interviews were unusually long and becoming less relevant. When we started analysing them more closely, we changed our minds.

Sharing the research journey with older people who would traditionally only be invited to take on the roles of interviewee or questionnaire respondent, has been immensely rewarding and satisfying for us personally, as well as being beneficial for the research. It is hard to capture succinctly the effect collaborating in this way has had. Perhaps it is best to let the quality of the interviews drawn upon in this book speak for themselves.

3 Housing Decisions in Later Life

Introduction

Moving house is recognized as one of the most stressful events in people's lives. In the main, the assumption is that the stress is due to the factors related to relocation: deciding what to take to the new house; arranging removals; having to end and start services, and tell people about change of address; sorting out the new house, including building works; getting used to living in a new area, getting to know people and to sort out daily living arrangements. All of this can be, and frequently is, highly stressful.

However, there is another aspect of house moving that is more significant for this study: the process of choosing where to live. The factors that are taken into account in determining where to go will differ between people and life stages. Obvious elements are size and type of house, locality, facilities and access to the house and within it.

In this chapter we examine the context in which older people decide where to live. We recognize that in doing so we are making an artificial split between deciding whether to move and where to move. In practice, decision-making is not as neat as this. People may decide the time is right for a move, but then struggle to find a place that meets their requirements. So they change their mind. The process of looking for a house may lead to reflections on what is really wanted. In turn this may spur people to narrow their search or abandon it as they realize they are better off where they are. Alternatively, a decision to move is made, but circumstances change and the decision is reviewed. In housing decisions, whether and where to move are intermixed.

A central and recurring theme in social gerontology is the obvious, though frequently ignored, essential that older adults are the same as younger adults and yet that there are features that are distinctive to being older. We shall draw attention to factors that are common to all age groups but will give more attention here to what is special about housing decisions in later life.

Moving in later life

Characteristics of moving after 60

Two interviewees gave accounts of their housing moves:

> We had a large Victorian house. ... And I looked at the house, I used to do the decorating ... and I thought there comes a time I will not be able to do these ceilings. ... And I'm looking in the local paper one day, I said to Neil, 'Oh that house looks a nice three bedroomed semi'. He said, 'Phone up the agent and go and see it'. ... Anyway I phoned up the agent and he said, 'That's gone but I've got another one coming on the market'. (Mrs Keeley, 71 years, married, own house, London)

In due course they found a house that they bought. Elsewhere in the interview Mrs Keeley explained the reasons: (1) they foresaw problems with decorating; (2) they wanted to live in a smaller house that would be cheaper to run; (3) they had known the area they moved to for some time; and (4) it was on the tube line – for 'the convenience of travel really'.

Mr and Mrs Carpenter had moved to a flat 6 months previously from a house that they had lived in for 40 years. They had decided to move because of their health and their age.

> Because my husband's health is not good and our age, we are in our 80s now. So because the house and the garden were just getting too much. ... Well, you see, we, unfortunately, we haven't any children ... we just rely on each other. And Roberts House had a special invitation way back in the early part of the year and we came here and saw the place and we were quite impressed with it, and everybody made us so welcome that we thought this was a nice sort of area to come to, and that's the reason why we chose to come here. (Mr and Mrs Carpenter, 3-bedroomed flat, housing association, London)

Several features emerge that are distinctive to moves in later life. First, some people make preparatory moves, ones where they are getting ready for the future. Thus Mrs Keeley thought that they would have problems decorating and saw the advantages of somewhere that was cheaper to run. Finance is a major anxiety for people on fixed or declining incomes: some want to be able to live within what they see as good enough standards, for example, being able to buy presents for grandchildren and ensuring that they can pay to heat and repair the house; others want to be in a position to negotiate housing and care whatever arises.

The second example is of people moving to deal with current problems: Mr Carpenter's health, managing the garden and their age. Advancing age is presented in this account as a problem that does not need to be defined. Capacity to manage in older age is referred to obliquely with the reference to not having children: the implication is that those without children lack a resource available to others. The particular flat was chosen following a process of getting on the lists of different housing associations and then liking the place when they visited.

When people decide where they want to live, they take account of the house itself and its place in the locality. They want both a place in which to live and a home that they create. Of course what they look for will be influenced by a number of factors: their experience of the houses they have lived in, their own state of health and that of others who live with them, the money they have available and changes in their lives. It is not only that their health gets worse or the place becomes run down: it is also that their priorities may change. In this way a long-term preference for an old house or a large garden is influenced by current circumstances. The housing preference may be transformed into 'a house where I can express myself' or 'a conservatory for indoor gardening'. At times the preference seems diametrically opposed to an earlier objective. People who have loved gardening may prefer to have no garden at all if they are unable to work the garden as they once did, though they may still want to have a room where they can look out at trees and see the changing seasons.

Preferences are influenced also by personality and decision-making style. Thus, some who are fit will want a smaller place on the ground floor so that they are ready for the possibility of becoming less able and finding the stairs difficult. Others, by contrast, who are already finding the stairs difficult, choose not to move because they want to stay in a home that they love for as long as possible. They may or may not have thought about the problems that may emerge in finding a suitable place when they have less energy and mobility. In effect, they may well be making a clear choice not to make changes for the time being but to react to what may happen. This type of thinking about circumstances and reaching a decision to react later, is very different from those who cannot face any thinking about what to do: ostrich like, they bury their heads in the sand. The complexity is explored in more detail in Chapter 7.

A major part of the dilemma as to where to live in old age arises from the impossibility of predicting the future. So it becomes very difficult to identify and prioritize the range of current and future events which may have an impact on housing and management of daily living. Some of these factors relate to circumstances that can be relatively stable; others are more likely to be volatile. Neighbourhood facilities and finance are factors that have less propensity than others to change. Health and family relationships may be stable for long periods but may change to become volatile. These all provide the structural context in which housing decisions are made. Housing decisions may be an attempt to deal with changes in structural conditions. Further, whether people have developed, or are able to develop, tactics, strategies and routines to cope, clearly affects both the process and the outcomes of their housing decisions.

To stay or move?

Some people decide to move and others stay where they are. Moves are more likely in early old age or later when in their eighties (Rolfe *et al.*, 1993). The explanation for this seems to be that some use the change in their lives consequent on retirement both to take stock of where they are living and to plan for the future. If elders do not move at that time they are less likely to move unless their circumstances change. Changes may be of the type that make people aware that their situation is less stable than they had thought, for example, if family members move from the locality or a partner dies. A health problem which is not in itself disabling may alert the person to the volatility of their setting. Finally, there are those who find themselves in a crisis so that coping in the place where they live is impossible. Heywood and colleagues add further points: 'as people get older in Britain, they are less likely to move home'; '73 per cent of those aged 65 or more had not moved for ten or more years'; most moves by older people 'were made by owner occupiers moving into alternative owner occupied properties'; and, in terms of geography, they were least likely to move into large urban areas (Heywood *et al.*, 2002, pp. 78–9).

There has been some consideration of the reasons why others choose to stay. There are the obvious gains of continuity of place and life style as well as the dominant perspective that, until the situation becomes intolerable, it is best to stay where you are. For some the prime reason for staying where they are is that their picture of themselves, and other people's pictures of them, is best maintained in their house. The problem with what is seen as a forced move is that the person is thought to have had their control reduced. One comment on our findings has been: 'Space is complicated. It is about your place in society. There are occasions when you are placed – so space may be associated with lack of status' (True for us? 2003). Hepworth (2000) writes: 'We usually think of people in terms of the places they inhabit or are located' and that, 'Location is closely linked through categorization to social identity' (p. 86).

Oldman (1991) contends that 'Older people's main reason for not moving was the perception that they could not afford to do so' (p. 261). If this is the case, there is a probability that there are numbers of older people living in accommodation in which they find it difficult to manage. Heywood and colleagues cite other reasons for not moving:

the reticence or inability to expend the physical and mental energy required to undertake a move;
community ties and social networks;
material culture and attachment to home;
enforced decision-making in a confining relationship;
lack of available alternative housing; and
perceived suitability of alternative housing (Heywood *et al.*, 2002, p. 81).

Means (1999, p. 305) cites two main housing problems for those in early old age: availability and affordability. Those of the respondents who were considering a move nearly all wanted either specialist housing or ordinary housing which was easy to live in and maintain. For most this meant a bungalow, or a flat with a lift. Many found it difficult to locate accommodation in the area they wanted. A move such as Mrs Keeley's to a semi-detached house was comparatively unusual.

Quoting from a 1996 survey, Means notes a more general housing problem that is relevant to housing decisions, the condition of the property: one in five of all dwellings had urgent repair costs of more than £1000 and poor conditions are found across all sectors. The private rented sector had the greatest proportion of unfit homes (19.3 per cent) compared with 6.3 per cent of owner occupied property and 5.2 per cent of housing association stock. 'People on low incomes are the most likely to live in poor housing conditions and this includes many older people especially after the age of 75' (Means, 1999, p. 307).

Respondents to the questionnaire in the *Housing Decisions in Old Age* research showed a remarkable satisfaction with their housing. Ninety per cent thought their present homes were suitable for their current needs. Nearly three-quarters of owner-occupiers considered the amount of space in their present home to be 'about right'. Seventeen per cent of respondents were finding stairs difficult to manage within their homes and a further 2 per cent had installed a stair lift to cope with their own or their partner's difficulties with stairs. Surprisingly, over three-quarters of the owner-occupiers were completely satisfied with the internal and external condition of their houses. In this survey 75.1 per cent were owner-occupiers.

Such statements of satisfaction should be put in context. It is possible that 'owner occupiers might be less likely to express dissatisfaction with the appearance of their homes than tenants' because of the fact that they know they are responsible for the place (Kellaher, 2002, pp. 50–1, quoting from the General Household Survey). Further, Oldman (2000) in a qualitative study 'showed the disabling effect of much of the housing occupied, and that a significant number of owner-occupiers wanted to leave the tenure' (Heywood *et al.*, 2002, p. 81).

Appleton makes the point that the reasons that individuals may stay where they are may not be because of the positives of the current house: not moving 'often reflects either an absence of attractive alternatives or a lack of information about the alternatives that might be available' (2002, p. 13). This fits with our own finding: over a third of the 563 questionnaire respondents said either they 'definitely would' or 'might' consider sheltered housing *immediately*, if it were available in their local neighbourhood.

Moving in advance of a crisis differs from moves when younger in that, typically, the move is not necessitated by an event such as a change of job. The fact of there not being a pre-determined time scale makes the decision different in that it does not have to be made by a set date. A second group move when they come to recognize that their situation is volatile, perhaps following a fall, family moving away or the death of a partner. The third group is made up of those who are in a crisis and have realized that they cannot stay in their current house.

In planning a house move in later life, people are trying to take account of current and prospective changes in health and ability: 'Will the place be suitable for me/us if ...?'. Of course, a part of the complexity is that we know neither our future health nor the life events of people close to us. Some who may be contemplating moving to be closer to their children will also try to assess the significance and stability of relationships with family members other than their partners. Here again there are uncertainties: what if their adult children move to another area or split up? There are also the obvious difficulties of assessing what will be the impact of living closer to particular family members.

Theorizing ageing

There is a tendency to think of moving in later life as a response to problems, part of a perspective that regards old age as a time of decline rather than as a continuing journey. By contrast, for some, such house moves are an indication of wanting, positively, to change lifestyle in ways which a new house may make possible. Hepworth has some powerful examples from his study of fictional accounts of ageing. Commenting on Lady Slane's move of house in Sackville West's novel *All Passion Spent,* he writes:

> The move of place signifies a change of self – pleasing oneself in later life ... Her move is a declaration of independence by an older woman from the domestic sphere. (p. 88)

One way of looking at what is wanted in housing moves is to consider this type of distinction between those who move to deal with problems and those who move because they want to establish something new in their lives. Such a distinction has its uses, and it would be foolish to deny the fact that, for most older people, dealing with the negatives is a more powerful force than creating new opportunities. However, the split between negative and positive factors hides the reality that is captured in the picture of life as a journey. Body decline is a part of ageing, though obviously in different ways and with differing impacts. Yet, the changes brought on through ageing, the problems encountered on a journey, can lead to new opportunities. Lady Slane moves following the death of her husband. In our study, a woman described moving to a smaller home and using the event to get rid of possessions: she knew better what were her essentials. What might be termed 'positive' and 'negative' intertwine.

To locate our study in wider theories of ageing we pick out three perspectives that have particular relevance.

Life course

A stage is reached at which people become, or are defined, as 'old'. There are changes to people's customary physical activity in old age whether as a consequence of health or circumstances. Attempts to promote a more positive perception of old age may fall into the trap of disguising the real changes and losses that take place. Yet in recognizing the reality of the changes, typically changes which are unwanted, we must ignore neither the capacities of older people to adjust nor people's differing responses to such changes.

Individuals can be described in relation to the experience of ageing. Some will be described as battling against the decline, as good for their age or as ageing prematurely. In effect older people are seen as people with something missing or they are lauded because they transcend the stereotype, for example, by taking up diving in their eighties: such an approach has been referred to as 'deficit' and 'heroic' models of old age (O'Neill, 2003).[1] With such perspectives, the normality of ageing is ignored, and with it the majority of older people. Blakemore (1993) proposes three types of ageing amongst ethnic minorities in the UK: a self-reliant pioneer; a gradually adjusting response; and a passive victim.

One of the repeated themes in writing about old age is that of people distinguishing their inner selves, the 'real me', from their body images: 'Inside', they will say, 'I feel young'. How are we to understand the experience of being old – and of being ourselves? Do we have a continuing or a changing identity?

Life-span developmental psychology stresses the importance of seeing 'people in the context of their whole life history'; there are reciprocal influences between person and lifespan (Bond *et al.*, 1993, p. 30). Hepworth (2002) states that he is trying 'to reflect the interplay between the body, the self and society' (p. 9). He draws on symbolic interactionism, describing it as:

> one of the branches of sociology that places a high value on the role of the imagination in the development of the concept of the self. According to symbolic interactionism our sense of individual selfhood develops from infancy through the human capacity to become aware of the way others see us. (p. 6)

Another strand in the cloth of ageing is that of thinking about a person's life course, career, biography, project or journey. All are terms used to portray a number of factors: the interaction between people and the events of their life; the way people reflect on, manage and use the events; and the construction of the stories people tell themselves and others about their lives. The ways in which we tell our stories convey something of our sense of being in the world, that is our relationship to the world. In the process different perspectives on time emerge.

Arber and Evandrou (1993) analyse components of the life experience of different groups at particular stages of their lives: older people in relation to their history and culture; and the interaction between the passage of individual time, family time and historical time. Time is one of the hardest concepts to define in relation to ageing, but is important in that housing moves take account of perceptions of time.

Chronological age, as Hepworth (2000) notes, is a common conversation topic that is used to establish an 'age identity'. 'Time perspectives' are employed:

> References to experiences which have taken place in the past; places lived in and visited in the past; associating the self with the past; and drawing attention to change … . (p. 57)

'Reflection on how time has been used' is a theme examined by Biggs (1993, p. 39). 'Looking forward, it is necessary to anticipate death, to take it into account as a boundary to one's plans.' He writes of looking back and looking forward, recognizing that the time perspective changes 'from time lived since birth to time remaining until death' (Biggs, 1993, p. 41, referring to Schroots and Birren, 1990). Awareness of the imminence of death, as he points out, 'may also change attitudes to time across the lifespan' (p. 42). He considers the shifting relative dominance in people's time frames of past, present and future. This perspective is illustrated by a study into quality of life:

> Past experiences, current and future prospects, and considerations about other paths that their lives might have taken were all used as ways of setting markers for assessing quality of life. (Butt *et al.*, 2003, p. 1)

In addition, time can be seen in the present in relation to the way time is used: people speak of 'free time', 'personal time', 'filling time' or 'filling in time' and 'wanting more time to …'. All of these phrases gather significance as people realize that the time left is limited. Victor and colleagues (2003) in a study of loneliness and social isolation examined the amount of time spent alone. They note that although, as is obvious, being alone and being lonely are very different, increased time alone was one characteristic in increased vulnerability to loneliness.

Recognizing that death is not far away is one of the realities of ageing. Inevitably this is a central feature of people's final years. Many writers see old age as a time of reflection. Erikson's (1950) model highlights various stages in life and outlines 'tasks' that have to be fulfilled. In old age the tension is between ego integrity and despair: do I see myself as whole? have I done what I wanted? am I satisfied with my life? The focus could be seen to be a combination of who we are and where we are going. Gandhi (2002) writes of the perceptions of South Asian elders in London of later life:

> It is a time for spiritual reconciliation, a time for preparation for death and a time for taking stock about the course of one's life, fulfilling duties to one's family and friends and moving one step higher in personal self-achievement and self-evaluation. (p. 138)

Quality of life

'Quality of life' has become perhaps the dominant way of looking at whether life is satisfactory in older age. Earlier debates tried to locate 'successful ageing': was it to be found by replacing former work or family tasks with other activities, as promulgated by activity theorists, or by disengagement, in which individuals and society accepted a withdrawal from former roles? (See Havighurst, 1963; Cumming and Henry, 1961.)

Nazroo and colleagues (2003) compared ethnic groups on six factors that influenced the quality of life of older people: having a role, support networks, income and wealth, health, having time, and independence.

> Factors that are typically included in research concerned with inequality (material conditions, health, crime and physical environment and formal assessments of social participation) revealed a familiar pattern of great inequality. The white group tended to have the highest scores, followed by the Indian and Caribbean groups and then the Pakistani group, which had the lowest scores on each of these dimensions. However, for those influences concerned with less formal elements of the community – social support and perceptions of the quality of local amenities – differences were reversed, with older Pakistani people better off than older people in other ethnic groups. ... Of all the factors examined in the quantitative analysis, the gap between older people in the white and ethnic minority groups was largest for income and wealth and housing conditions, with women having lower scores overall in all four ethnic groups. (pp. 2–3)

Differences in life experiences and life chances of women and men are highlighted here, as well as those between ethnic groups. Another recent study by Beaumont and Kenealy (2003) has a different list of components:

> The most important factors in determining a perceived good quality of life (QoL) were the individual's perception of their health, freedom from depression, personal optimism, well-retained cognitive abilities and aspects of the social environment. The common themes concerning their QoL, mentioned by participants as important, were issues related to their family, their health, and to the conditions associated with their home. With respect to residence, those who were living with a partner tended to report the highest QoL; those in residential homes, irrespective of their health or disability, reported the poorest quality of life. (2003, p. 1)

Poor and worsening life satisfaction is most clearly associated with worsening ability to perform daily living tasks and with health status. Approaching one-third of older people at any one time have poor life satisfaction, though some people move to and from being satisfied (Bowling *et al.*, 1998). More recent work by Bowling and colleagues (2003) notes that the majority of people rate their quality of life as good. They show that how quality of life is rated depends on: comparisons made with others and with one's own expectations; a sense of optimism or pessimism; good health and physical functioning; engaging in a large number of social activities and feeling supported; living in a neighbourhood with good community facilities and services, including transport; and feeling safe in one's neighbourhood. Many of these factors match the views of participants in this study.

Three additional themes that have not yet been mentioned taken from a longer list of factors that formed the foundation for a good quality of life are:

- living in a good home and neighbourhood;
- having adequate income;
- maintaining independence and control over one's life. (2003, p. 1)

Ageing in society

Perceptions of ageing have immense significance for the ways people understand their own and others' experiences. A contrast is frequently made between Western and Asian or African attitudes to older people: although carrying the dangers of all stereotypes, this one holds a significant truth. Butt and colleagues (2003) report on the impact of ethnicity, noting that:

> ethnicity plays a part in older people's perception of their quality of life. In terms of minority ethnic older people, there may be a positive or protective element in that they have a more positive view of the process of ageing and the social support they can draw upon.

For example, some Muslims 'emphasized that there was a divine purpose behind what happened'. Further:

> Chinese or Asian participants were more likely to report that growing older had positive aspects, such as becoming wiser or more tolerant. Here, the strongest contrasts were with white women, some of whom expressed negative ideas about ageing, either in terms of how they felt about themselves or how others reacted to them.

Comments from three elders in focus groups bear out the perception 'that older people are not wanted in our society':

> No, because we are not wanted really.
> The National Health doesn't want us.
> No, society can't cope, this is the thing. (Wright, 1999, p. 247)

By contrast:

> For some minority ethnic women, especially Indian and Polish participants, there was a connection made between a sense of well being and being respected and valued by others. This was particularly in terms of the status afforded to older people in their cultures. (Afshar *et al.*, 2002, p. 2)

Hepworth uses the term 'ageing into old age' as a reminder that *old age* is a construct: it is defined by society, changes and has different meanings for different people (2000, p. 2). There are differences in how people perceive ageing. Afshar and colleagues report that some Pakistani and Bangladeshi women felt older at a much earlier age than other groups (2002, p. 2).

Social constructs, such as the role for an older person, impact on people's perceptions of self. But sense of self is influenced also by people's internal state, their sense of self worth. The sorts of houses that are thought appropriate for older citizens are determined by the perceptions held about older people's life style. And of course, individual older people consider how and where they live within constructs, of their group (whether defined by class, ethnic group, gender, religion or other affinity group) as well as of wider society in the UK. Not only may elders take account of these constructs, they may wish to live alongside others with whom they share fundamental beliefs.

Public and private worlds

The notion of one's house as a castle portrays a place of safety within and the potential to keep others out. Studies of old age homes from the 1980s onwards have examined public and private worlds. Clough (1981) wrote of home as a base, with a boundary between the inside and outside. He noted how this changed in a residential home where much of the daily routine, typically a private function, is carried out in the semi-public forum of a sitting room. How far could life styles and architecture of residential homes change, he asked, to re-create some of the characteristics of a base? (pp. 94–9).

One of the features of housing options for older people is that the private world diminishes and the public increases. Most age-specific housing (and all residential and nursing homes) have some communal features. As will be discussed in a later section, communal facilities and services are considered one of the distinguishing aspects of such housing. The earlier discussion of social constructs creates the framework for consideration of the reasons that the balance of public and private shifts. Is it because elders are seen as people who have reduced lives (and therefore need less space) and need looking after (and so must be subject to observation)? In some scenarios, elders are to be kept cocooned, safe from harm. Consideration of how they are to live, in contrast to how they are to be looked after, may not be asked. Another dimension is what people perceive as their public and private selves.

Dominant ideologies and frameworks

Three core assumptions have underpinned recent welfare and housing policy for older people:

people want to live at home;
people do not want to live in residential homes; and
people want independence.

Each of these assumptions relates to the others but each has a life of its own.

The frequently asserted statement that older people want to live in their own homes is not surprising. Participants in this study reinforced this stance. However, as with any other such certainty of the time, it should not be accepted as 'truth' and simply re-asserted. 'Home' is the place where individuals establish themselves. In a later chapter we look further at the attachment to home but it is sufficient to note here that ideas of ownership, status and becoming an adult citizen all intermingle. Having lived in their own homes (of whatever type) it is natural that, when asked in old age about housing preferences, people assert that they want to stay in their own homes. Such views are influenced by the extent to which they know about other places to live in their area and the accuracy of that knowledge.

Heywood and colleagues claim that 'in Britain, in contrast to other European countries, there is widespread fear of the residential home, of the relinquishing of householder status' (2002, p. 9). They argue that this fear:

> is part of what drives people to remain in their own home against all odds, and to put up with discomfort, cold, loneliness and worry rather than to surrender to what is commonly regarded as the last resort.

Dalley agrees, referring to 'the pitiful plight of many people living in their own homes' (2002, p. 23). Clough has argued that 'the home is best' argument is often put by those 'who are not in a situation where they have to make a choice about living arrangements in old age' (1981, p. 13). He claimed that the reality of some people's lives in their own homes is isolation and loneliness. Against the dominant perspective of 'live at home', a few have continued to question the quality of life for many who do live at home. (For more recent examples, see Oldman and Quilgars, 1999.) Dalley asks whether people have life outside the home (2002, p. 16). Further, she argues that:

> ... in elevating the principle of independence as the essential virtue to be established and maintained at all costs in late old age, gerontologists and health and social care practitioners are misreading the 'lived experience' of many older people. (p. 18)

The fact that most older people do not want to live in residential homes is equally unsurprising. The residential provision of the nineteenth and early twentieth centuries took the form of the workhouse, for many encompassing the harshness of state provision. Thus the fear of not being able to look after yourself and live at home became translated into the fear of the workhouse. This public perception of harsh living was reinforced in the 1960s by academic studies that described the lifestyle within large residential establishments as geared to the interests of the institution, not the individual who lived there. The term 'institutionalization' came into being, and became the focus of many investigations and research inquiries in homes for older people and hospitals for people with learning disabilities and mental health problems.

The dominant and persistent image of residential living is negative, a perspective that has immense significance not only for those who live and work in homes but also for prospective residents (see Clough, 2000, pp. 46–53). Parker (1988) summarized this persistent image:

> The idea of institutional life has always been viewed with repugnance by a broad section of the population. This attitude has persisted despite many changes and improvements and although now it may be weakening, it nevertheless continues to be influential. Its survival has been assured by at least four forms of reinforcement: the deliberate cultivation of a repellent image; reported cases of the abuse of inmates; the enforced association and routine of institutional life, and the compulsion often associated with entry as well as with subsequent detention. (p. 8)

Some people may hold similar views of any specialist housing for older people, seeing at both as an indication of loss of status and a potential infringement of personal space. By contrast, several interviewees in this study were delighted with their moves. Here, Mrs Denson speaks warmly of sheltered housing.

Mrs Denson's story

I am, very, very lucky. I realize that. I, you know, I thank God every night that it will stay as it is. It is wonderful, really wonderful. I think a lot of the other people downstairs would tell you the same thing. I came here in December last year. I've been here about eighteen months now and it's the best eighteen months I've ever had in my life, it really is, it's wonderful.

I: Perhaps your experiences will help other people make informed choices about housing decisions?

> *Mrs Denson*: I hope it would, you know, because I knew nothing about any of this sort of thing. When the Doctor said to me, 'We think we'd like you to go to sheltered accommodation', I was horrified, I really was, you know. I said 'Oh, I'm not ready for that yet'. I don't know what my version of 'that' really was, I'd never sort of talked about going into sheltered accommodation, never wanted to. I was happy where I was. (Mrs Denson, 81 years, widow, sheltered housing, Lancashire)

Clough has developed the term *institutional tendencies* to describe the 'processes which can be countered but have the potential to dominate life in a residential home':

for people to lose or to be stripped of their identity when they move in;
for residents to feel powerless;
for a person's privacy to be invaded;
for residents and workers to seem to be on opposite sides, as 'us' and 'them';
for it to be easier to care for people in groups than separately;
for (some workers) to find the task beyond them and end up controlling residents in ways which they would have found unacceptable when they started (Clough, 1993, pp. 79–80).

Oldman and Quilgars (1999) comment on the development of research and policy-making. They argue that the body of work relating to living at home/in a home is dominated by a 'structural dependency paradigm' where 'home' is seen as embodying personal control and self-identity, and residential care regarded as exemplar of 'institution', with all its connotations of dependency and passivity. Institutionalization, they argue, is endemic to the lives of some older people, regardless of where they live, seen, for example, in some highly routinized services in people's homes; indeed, they contend that the delivery of services may exacerbate an older person's dependent status.

Their research, which included an investigation of residents' lives before moving into a residential home, uncovered positive experiences of residential life. These went against many of the stereotypes associated with being in a home in that, for example, residents often made their own decisions to move into a residential home and were not necessarily isolated from the wider community. They argue that studies of many older people's situations *before* moving reveal loneliness, which is a further factor in the demand from older people to live in settings where there is closer contact with others.

Nevertheless, in trying to understand people's housing decisions it is essential to recognize that, whether or not they are correct, there are deeply held views of the poverty of residential life and, to a lesser extent, of other specialist housing. The debate about the quality of residential homes is not the subject of this book though we stress that we do not endorse the view that residential homes are either unnecessary or, inevitably, offer second-rate living.

There is interplay between preference for living at home and fear of a move to a residential home. In the process 'a stereotypical picture of the poverty of residential life ... is contrasted with a picture of perfect life in one's own home' (Clough, 1981, p. 11; 2000, p. 52). Not surprisingly, people who are facing significant difficulties in managing their daily lives are more likely to think a move to a residential home to be a good thing.

The emphasis on living at home may lead people to consider any alternative to be a failure. Thus 'ageing in place' has become the favoured approach both of older people themselves and of governmental policy (Klein, 1994), and there is a great fear that moving will lead to a loss of independence (Bland, 1999).

Independence is the third core assumption we noted at the start of this section. Once again there is cohesion between individuals' statements and those of governments. It has become commonplace for ministers to assert independence as a test of policy direction and outcome. Indeed grants have been allocated to services that 'promote independence'. This leaves an uncertainty as to whether people who are becoming more dependent, perhaps because of the nature of an illness, will be seen as less important. Individuals' perceptions of their own ageing are influenced by these dominant ideas. Repeatedly when asked what they want in old age, people list independence as a key factor.

The problem with such dominant ideas is that they may be used with little precision. In old age many people find ways to manage increased incapacity and dependence. So a mechanism or a person may be found to help manage a task that has become difficult; alongside this someone may take up exercises as a means of strengthening muscles. Second, tasks may be given up: people may stop doing certain gardening activities, perhaps adapting the garden or getting someone else to come and mow the lawn. How is the term 'independence' to be understood in this context?

> The points seem almost too obvious to need to be made: we move in and out of dependence throughout our lives; sometimes this leads to us managing tasks without the support of others for most of our

> lives; sometimes we need others to help us with certain tasks most of the time; sometimes we do tasks for ourselves and sometimes want (and sometimes need) others to do them for us; this can vary with our physical and our emotional state. (Clough, 2002, p. 34)

Wilkin (1990) has examined the term 'dependency', noting different causes: life cycle, dependency of crisis, disablement, personality trait, socially/culturally defined. These ideas are useful because they shift the discussion from an over simple analysis looking at an individual and how the person is treated to the impact of other factors on people's lives. For example, people may be dependent on others not because of their own lack of ability but because poor transport means that they cannot get to the shops or travel to see their family.

Similarly, many advocates for 'staying put' argue that the focus should be on disabling environments rather than people with disabilities: independence-promoting environments improve the level of functionality of older people. Thus, altering existing housing to a better 'person-environment fit' would prevent unnecessary moves, often related to fears of injuries and to the difficulties associated with daily living activities (Klein, 1998). This fits with the wider critique of models of disability, in which, at its most simple, a medical model, seeing the problems in the individual, is contrasted with a social model, in which the problems are located in the attitudes of society which fails to provide adequate access, transport and so on.

Similarly isolation, problems in cooking or poverty (all key potential factors in a state of 'being dependent') may result from poor housing: steps that prevent people with mobility difficulties from getting out, badly designed cupboards or cooking equipment that limit what people can do, or expensive heating, coupled with poor insulation, that results in cold houses or too much money spent on heating.

The discussion of independence has a direct bearing on housing aspirations. Under the banner of 'independence' it may be that people are looking for 'control of lifestyle' (Clough, 1981, pp. 31–2) or involvement in decisions that affect their lives. Indeed, rather than repeating as a mantra, slogans such as 'housing for independence', we may need to understand more of what people want from living in their own homes.

Dalley contends that the 'aggressive assertion of autonomy' as an objective may get in the way of those coping with the onset of infirmity: there may be more fruitful ways for them 'to come to terms with their day-to-day experience'. She cites 'counterbalancing concepts':

> Companionship and friendship; mutuality; common interest; support and personal care; social activity; the easing of responsibility for daily functioning ...; freedom from fear of isolation, loneliness and danger. (p. 19)

Housing in later life

Perceptions of older people have had dramatic impact on housing assumptions. It is presumed that old age is a time of reduced physical capacity and, therefore, that older people should have reduced housing. Nearly all specialist housing for older people has fewer rooms and, just as important, smaller rooms than standard housing. A second feature of specialist housing for older people is the addition of communal facilities. These may include: a restaurant, a hairdressing salon, a bathroom with special hoists, meeting and activity rooms, laundry rooms and guest rooms. A third key strand is the extent to which there are services available to support people. Put in this way, the terminology of 'available services' is neutral. However, it is common to regard the existence of services as meaning that older people need to be 'looked after'. This is the route that ends with the word 'care' being demeaned to the status of the opposite of independent living. A move into 'residential care' is assumed to mean loss of control.

Thus, housing in later life can be divided into:

general housing: older people living in general housing, which may be more or less suitable; the *lifetime* home movement has wanted to see all housing built to high standards, which would be suitable as it stands or would be capable of adaptations for people through out their lives; this includes all forms of tenure;

age-specific housing (or *specialist housing*): what is distinctive is that the houses are available only to older people; thus there are numbers of older people living alongside each other; more modern schemes are likely to be built either on the ground floor or in buildings with lifts and to be accessible to people in wheel chairs; they will have some communal facilities, which may be available to others in the local area; there will be a housing manager and, probably, there will be a system to call for help in an emergency; there may or may not be services available; available in all forms of tenure;

residential home/nursing home: nearly everybody will be assessed as to suitability before moving in; 'residents' are neither owners nor tenants; the private space is limited usually to a bedroom, in more modern homes with toilet and bath/or shower facilities; staff are on duty 24 hours; meals are provided and assistance is available for all daily living tasks.

The amount of space and the way it is organized is a theme that is central to this study. Many older respondents are angry at the limited amount of space that is provided in specialist housing. We share their concerns and argue in the final chapter for far higher standards in specialist housing. Nevertheless, it is necessary to recognize that there have been trends in general housing, perhaps from the 1960s or 1970s, for there to be an increase in the number of rooms, particularly bedrooms, but for there to be large reductions in the size of rooms. Indeed the popularity of TV programmes telling people how to get rid of their possessions no doubt is in part influenced by the fact that storage in general housing is very limited. People have to think of innovative solutions; there were recent reports of families with very little storage space having to store goods in warehouses on a permanent basis – they collected and returned items on their way to and from work.

Within these very broad groupings there are numerous variations. In particular age-specific housing can be either very similar to general housing, with very few services available, or, in the form of extra care or very sheltered housing, can provide services similar to those in a residential home. The distinguishing characteristics between general and age-specific housing are that in the latter, first, only elders are eligible and, second, there are some communal services. Specialist housing and residential homes are distinguished by differences in tenure. In residential homes, there are additional features: 24-hour staff cover; the availability of more services; people have private bedrooms or bed-sitting rooms in a building rather than separate flats or houses.

The life styles within these different types of housing are influenced by prevailing attitudes about the functions of the places, which impact on the people who live in the homes, staff and outsiders. Of course, there are significant influences also from the structure of buildings. Finally, the working style of staff has a dramatic effect on day-to-day living.

Research reports show that there are good and bad life styles in each of the major categories, though, as noted above, it seems presumed that general housing is good, age segregated housing may be good and a residential or nursing home is to be avoided. It is noticeable that residential and nursing homes are not typically regarded either as skilled specialist resources nor as good places to live and die, as is a hospice. Yet there are numerous contrary instances:

older people in general housing whose lives are very limited – lonely, isolated, depressed and deprived;
people who find that staff in specialist housing treat them as people who have to be organized; there are several examples of this in this research;
residents of residential homes who find companionship and enjoy living.

Most writers classify the types of housing, and then consider the sorts of services that are available. However, it would be possible to start with classifying the services in terms, for example, of:

the types of service offered: personal care (e.g. help with getting up or bathing); meals (help with cooking, provision of meal in own home, restaurant service); cleaning and house maintenance; gardening; advice; minor household tasks (e.g. changing light bulbs; repairing fuse);
the management of the scheme or service: private, voluntary organization or local authority;
payment levels and systems: rental, ownership, service charges; fixed, variable, annual;
involvement of individual tenant or owner in the selection of the service provider and the management of a service allocated to an individual;
the arrangements between housing organization and providers of services: is the same company responsible for the care services and the housing? *communication* between different organizations.

Two particular types of housing scheme illustrate the diversity of provision. Brenton (1998, 2002) writes about 'co-housing' which has been developed in Denmark. The hallmarks of the system are:

there are self-contained units of accommodation;
'members live on one specific site with the aim of sharing common activities and common space';
it is set up and run by its members;
the extent to which 'members live in common is a matter for choice';
'entry to the group is governed by a desire to live as part of the group' ... (Brenton, 2002, pp. 142–3).

One of the aspects of some forms of specialist housing is that people have no say over whom they live alongside. Of course this is the case with general housing but living with people with whom you may have nothing in common in situations where living arrangements have to be shared is a very different matter. In co-housing, individuals select the people with whom they would like to live and, at the beginning, move into the house as a group rather than individually as strangers. It is interesting that one of the driving forces for the Older Women's CoHousing Project Ltd is that individuals wanted 'to plan life together as a community' (Brenton, 2002, p. 141). Similarly Kellaher (2002) studied residents at Methodist homes: what made the places distinctive is that residents wanted to share experiences with other residents and staff and felt that they had things in common. Kellaher uses the term 'mutuality' to describe the life style.

A second type of specialist housing is a retirement village or community. One example is *Hartrigg Oaks*, a complex of 152 bungalows, a centre with communal facilities which functions also as a base for services, and a 42 bed residential home. The intention is to provide 'a continuing care retirement community' which enables people to stay living on the site up until the time they would have to move permanently out of the residential home. It also offers a choice of methods of payment based on a capital payment and an annual fee. The differences occur in the amount of risk carried by the individual: a higher fee means that care services, *if needed,* will be provided free; a lower fee results in charges for future care services.

Berryhill Retirement Village is a second example, with 148 flats, a shop, gym with jacuzzi and sauna, hair salon, computer room, library, craft room, woodwork room, licensed bar, village hall, cafe bar and restaurant. It aims 'to provide self-contained accommodation, responsive, individually tailored support packages, and ... on-site leisure facilities for 170 residents' (Berryhill, 2004). The model is one of activity and opportunity. Their publicity claims:

> The emphasis is firmly on raising aspirations, and challenging assumptions about age and disability: applicants have embraced the opportunity with gusto leaving no doubt that the young do not own the franchise on energy, self-expression and achievement! A determination to empower residents, by promoting user choice, control and involvement in their own environment, will typify every aspect of village life. (Berryhill, 2004)

The purpose of mentioning such schemes is to illustrate the diversity of specialist housing provision rather than to review their outcomes. However, some brief points germane to this study will be made from a recent study of Hartrigg Oaks by Croucher and colleagues (2003):

> residents were similar to most older people in citing as reasons for the move their worries about isolation, not wanting to pressure children, and health; however, they 'had generally been *anticipating* the implications of a deterioration in health' whereas others typically move when their health has deteriorated; (original italics, p. 7)
>
> the good reputation of the operating organization influenced many; (pp. 11–12)
>
> they hoped to be in a community of like-minded people; (pp. 12–13)
>
> they saw the move 'as an attractive insurance against two risks to independence, ... a decline in health and ... having one's life taken over by others as a result of a decline in health'. (p. 11)

Our study reinforces information from elsewhere that in specialist housing for older people there has been insufficient negotiation between planners, architects and the people who live in the buildings. Torrington (2002) outlines some of the problems. Architects are given as their brief a picture of older people with 'a great number of physical and possible cognitive impairments' who 'need to live in protected environments'. She recognizes that this analysis has some truth but concludes that the result is that people are seen as 'highly problematized with special needs'. 'This image is of a medical problem, not a person' (p. 193). She argues for greater involvement of older people to ensure that buildings are not sterile: life has to be experienced. For example, she states, buildings should have natural ventilation and light, not corridors which are permanently lit and show no change with day and night (p. 192). The design of buildings must take account of life style approaches: the objective should be to find a way of living and a building that best allows older people to live in ways they want.

However, it is important also to recognize the link between design of specialist housing and theories of ageing. If old age is seen as a period dominated by physical impairment and if it is assumed that older people have to be protected from risks, then it is not surprising that the end result is unexciting and sterile buildings. Risks can be minimized at the same time as recognizing that older people also want the adventure of living.

4
Understanding Housing Decisions

Four people comment on their housing moves.

Mr Harding
Mr Harding: Ilkley didn't suit us because of its position ... and so we looked for something else and had a definite plan of what we wanted. A fairly mild climate, a town of something like 15 000 inhabitants to suit the shopping and definitely easy access, preferably on a main line because of our children. We have to fit in with a daughter in London and a son in Barcelona.
I: So it was a major planning exercise ... you really had to take all of these factors into consideration, major planning?
Mr Harding: Yes ... and we considered a few places like Berwick, Hexham, Carlisle, Preston, and then Lancaster attracted us and we did a preliminary visit and confirmed we liked the position and the layout of the town. (Mr Harding, 94 years, widowed, 2-bedroomed flat, Lancashire)

Mr Saunders
One point we should make, that I feel very strongly, is that you've got to make a very positive decision that you are going to move into a place like this. You've got to analyse the situation properly and decide. You've got to do the two lists – the benefits and the costs. (Mr Saunders, married, sheltered housing, Lancashire)

Miss Davidson
I'd settled myself that I would probably be at Preston Patrick until I was about 70 because I just thought I wouldn't have sufficient points. I'm far too young and it will never be my turn ... and just

the prospect of coming home to Kirkby Lonsdale .... I thought, 'Well I've got to go for it, it could be years before I'm offered it again'. So that was it. It was probably a bit of a hasty decision. (Miss Davidson, 60 years, single, living in council owned, sheltered housing 2-bedroomed bungalow, Cumbria)

Mrs Panton

I've always believed that if you are in the right place, the time to move comes at the right time. I don't like indecision. But I think it was a learning period, a learning period to get to know myself really. So it is a long process of learning about myself I think. And my daughter had a flat here, a holiday flat. And I was staying with her and in a magazine there was an advertisement for a room to let in *The Dales*. And I didn't even know *The Dales* was a *Chadwick Housing* place. I didn't know much about *Chadwick Housing* at all. But I thought I'd go and look at it and put my name down if it's suitable, which was all I intended doing. And I came into this room – it didn't even have a light in the centre – and I looked out of the windows and something said, 'This is right'. And I always know when it's right, I don't have any hesitation. (Mrs Panton, 83-year-old widow, residential home, Cumbria)

All of these people have made decisions to move in their retirement. Furthermore, in each case, the people concerned have ended up happy with the results of their decision. However, what is striking about these statements is the different ways in which these people have made their decisions. In the first example, Mr Harding had a definite set of criteria (involving town size, climate, and easy access to children and public transport) which he used with his wife to determine what would be the ideal location. Mr Saunders also advocated the use of lists in deciding whether the advantages outweighed the disadvantages of moving. These men both represent a rational form of decision-making where their decision was based upon extensive research, long-term planning and calculating whether alternatives meet a set of desirable criteria.

In comparison, Mrs Panton relies on two things in determining that she had found the right place: her self-understanding ('getting to know myself' as she put it) and her intuition. In a similar way, Miss Davidson took advantage of an opportunity to move to a place that she knew she wanted to be. Her decision to move appears intuitive, motivated by her emotions. Again, she was able to seize her opportunities as they presented themselves because she knew what it was that she really wanted.

These examples provide immediate contrasts in approach: a more obviously rational decision-making style versus one that appears to be more influenced by emotions. In this chapter we uncover the tensions and debates about how people *are thought* to make housing decisions and contrast these explanations with our exploration of how they *do* make decisions.

Overview

The ways older people make housing decisions are complex. To add to the difficulty of unravelling this complexity, people may not be able to explain either exactly why they make the decisions they do, or whether their decisions are the right ones for them. They may be able to articulate some influences but not all. As we have been arguing, decisions are not necessarily the result of conscious deliberation or rational, analytical and objective thinking. It is not our intention to minimize the importance of a rational approach to decision-making. We examine its limitations and argue that its dominance has meant that the contribution of other ways of knowing are often hidden, for instance, when people think intuitively, act impulsively, or base their decisions on factors that are essentially emotional. In making decisions, people use two types of beliefs and knowledge systems: explicit and tacit, that is gut feelings, hunches and common sense. They also make their own subjective evaluations of the risks and benefits of certain courses of action.

Before examining decision-making theory, we summarize some central points from the last chapter. Deciding where to live in older age has similarities to deciding where to live at any age: there is an element of prioritizing the advantages and disadvantages of different options, trading off one good feature or aspect against another and, perhaps, having to decide where to compromise. Yet there are differences, because ageing is a time of changing physical capabilities. Planning and thinking ahead is difficult because it is not possible to predict one's own, or one's partner's, health and future physical limitations. There is evidence that there is greater difficulty if moving in crisis, in terms, first of participating and managing the move and, second, of creating the life style that is wanted after the move, for example in meeting others or getting support. There can be little doubt that, if individuals have already established their own routines and networks, they will manage better in any crisis. The same holds true for neighbours and communities: they are likely to be more supportive if they have got to know people before a crisis.

The thrust of current policy and practice appears in line with the preferences of individuals. However, dominant belief systems enfold individual older people as well as policy-makers. Two processes are at work: what individuals want is shaped by the assertion of social and cultural beliefs and values; at the same time, policies are influenced by what individuals state that they want.

In this chapter, we review the literature on theories, models and approaches to decision-making, and then outline alternative perspectives and research directions. The following three chapters will illustrate and develop these themes from the data. In the next section we focus on theories of decision-making as the backdrop to our later attempt to understand older people's housing decision-making.

Theories of decision-making

Psychologists have tried to understand the differences between the way we ought to make decisions and the way we actually do make decisions. They have looked at underlying thought processes to see which might be suitable for producing the best decisions. Theories of judgement and decision-making have started from the recognition that people have limited capacity for mental work and so, in order to make decisions in the real world, a number of simple ways of reasoning are developed which can lead to errors and biases.

Rational decision theorists suggest that a systematic and calculating approach should be adopted which allows the best option to be chosen from all the alternatives that are on offer. Bazerman (1994) sets out the following six steps:

1. Define the problem, characterizing the general purpose of your decision.
2. Identify the criteria, specifying the goals or objectives that you want to be able to accomplish.
3. Weight the criteria, deciding the relative importance of the goals.
4. Generate alternatives, identifying possible courses of action that might accomplish your various goals.
5. Rate each alternative on each criterion, assessing the extent to which each action would accomplish each goal.
6. Compute the optimal decision, evaluating each alternative by multiplying the expected effectiveness of each alternative with respect to a criterion, times the weight of the criterion, then adding up the expected value of the alternative with respect to all criteria.

This approach, it is argued, leads to the selection of the option with the highest expected value and a decision based on calculation, not on subjective, emotional reactions.

There is now considerable evidence to suggest that people do not think according to such rational principles (Baron, 1994). The main reasons put forward are:

- that people ascribe their own values to certain factors;
- there is a general failure to think through all aspects of a decision; and
- people use specific shortcuts.

Kahneman and Tversky (1979) in their *prospect theory* suggest that people are cautious about obtaining gains, preferring sure things to a risk. However, when faced with a situation where they will certainly lose, they may take a gamble that could lead to a bigger loss. Critics of prospect theory have examined the ways in which people actually simplify and reduce complex decisions into manageable ones (e.g., see Schneider, 1992; Wang, 1996). Another source of evidence for the conclusion that people do not follow rational principles is the observation that people fail to think through the consequences of uncertain alternatives (Shafir *et al.*, 1993). In particular, when people are faced with risky and uncertain situations they often use heuristic ways of thinking, short-cuts based on their own processing of data. Other important information that should be taken into account is overlooked (Gigerenzer *et al.*, 1999; Kahneman *et al.*, 1982).

Three examples of such heuristics are:

1. the *representative* heuristic;
2. the *availability* heuristic; and
3. the *anchoring-and-adjustment* heuristic. (Baron, 1994; Bazerman, 1998)

The representative heuristic leads people to make a judgement about the likelihood of an event, based on the similarity between that event and existing knowledge of similar occurrences. The availability heuristic refers to how easily past examples can be remembered, thus determining judgements about probability and frequency of events. Lastly, the anchoring-and-adjustment heuristic concerns the tendency for judgements to be biased because they are based upon the initial assessment of a situation arrived at with little or no reasoning. Decisions are affected by the way the options arising from a decision are presented: people are more likely to take risks when they think about what they may lose than when they think of what they may gain.

Decisions demand time, energy and the ability to focus on a problem. Simon (1956) suggests that people have developed short-cut strategies that deliver reasonable solutions to real-world problems, an idea that he referred to as 'bounded rationality'. Since then it has been argued that people avoid making difficult trade-offs between the good and bad points of an option. These are known as 'non-compensatory' strategies. For example, a 'satisfying' heuristic leads people to choose the first alternative that meets their minimum requirements (Simon, 1957). The advantage of such a strategy is that it does not involve as much thinking. However, it may result in less than best decisions being made, as potentially better alternatives are not explored.

Another strategy is termed the 'elimination-by-aspects' heuristic where decisions are made by identifying a main criterion and eliminating all options that fail to meet it. Subsequently, the next most important criterion is identified and, similarly, options eliminated which fail to meet it are eliminated (Tversky, 1972). Typically, experts making real-life decisions under conditions of time pressure and stress rarely consider more than one course of action at a time. Klein (1998) suggests that when making a decision, the consequences of following the same course of action that worked on a previous occasion are mentally rehearsed and only if this rehearsal is not acceptable will alternatives be considered. This is termed a 'recognition-primed' heuristic.

Having briefly outlined some general theories of human judgement and decision-making, we move to discuss particular theories and models that have been used to explain housing decision-making.

Explaining and predicting housing decisions

The last decade has seen more interest in the topic of housing decision-making, particularly in the US. Research in this field has a strong bias towards quantitative research and has tried to predict 'moving behaviour' by using theoretical models of decision-making. These models fall into various categories: micro-economic models (such as the human–capital model of migration and the cost–benefit model); socio-economic models and socio-psychological models (e.g., the life-cycle model, the stress-threshold model and the subjective expected utility model).

Micro-economic models

Human–capital models of migration have been used to explain mobility within the labour and housing markets. These labour-force migration models aim to determine the propensity to migrate based upon a number of different economic factors, but frequently conceptualize migration as an income maximization/optimization problem. In a similar vein, the *cost–benefit model* assumes that a person will only move if the benefits of the move outweigh the costs. In this model the present discounted value of all expected monetary and non-monetary benefits of alternative housing are compared with the benefits of the current housing. It is assumed people want to maximize their utility and will choose the housing where the gain of net benefits and costs is highest (DaVanzo, 1981a,b; Goodman, 1981; Speare, 1971 cited in Fokkema and Wissen, 1997).

Socio-economic and socio-psychological models

The *subjective expected utility model* predicts that people's determination to move will depend upon the sum of all the possible gains and losses of moving. Whilst it is similar to the cost–benefit model, it differs in its emphasis on the value placed by the individual on the consequence of moving and the expected likelihood that a given consequence will follow the move.

Vanderhardt (1995) has examined the socio-economic determinants of older homeowners' housing decisions. He found, again in an American context, that demographic factors might be more important than financial considerations in decisions about moving. Of particular importance were marital status, employment, children and disability. He applied what he termed a 'dynamic discrete econometric' technique to housing data on households whose heads were aged 50 years or older to estimate older householders' preferred living arrangements. This framework included:

- economic causes (e.g., home equity, financial assets, income, cost of housing);
- demographic/ psychological causes (e.g., housing characteristics, the home's psychological value); and
- health and social causes (e.g., physical limitations of health, familial considerations).

He found that a high level of home equity and lower levels of income encouraged movement, but that changes in demographic factors played a much more important role.

Johnson-Carroll and colleagues (1995) constructed a model of factors that influenced pre-retirees' propensity to move at retirement. They found that the older the respondent, the smaller his/her home, and the higher the maintenance skills of the respondent, the less likely they were to move at retirement. In addition, pre-retirees were judged less likely to move if they were healthy, looking forward to retirement, had spent many years in the community and had definite plans for retirement.

The *life-cycle model* presumes everyone's life follows a typical pattern: events such as getting married, starting a family, children leaving home and retiring from work will often lead to moving house (Speare, 1970 cited in Clapham *et al.*, 1993). Although the events are closely related to calendar age, it is not clear how the causal mechanisms operate (Thorns, 1985 cited in Clapham *et al.*, 1993). There are exceptions to this nuclear, traditional family of course, which do not fit the model: single parent families, couples who do not have children, people who never marry, those that re-marry and have second families, and those who divorce or separate (Forrest and Kememy, 1984, cited in Clapham, 1993).

In the *stress-threshold model*, the decision-making process has two phases. In the first phase, it is believed people will only move if their perceived stress exceeds a certain threshold level. Brown and Moore (1970) were amongst the original advocators of the stress-threshold model, defining stress as a discrepancy between actual and desired living conditions. Stressors can be *internal*, where there are changes in a person's needs and/or preferences. For example, a person, having become less mobile and unable to manage the stairs, might prefer to live in a bungalow, or someone unable to afford housing maintenance costs might move to a place with lower costs. *External stressors* are ones that occur outside the household, such as an increase in crime rate or in the noise level of neighbours. When dissatisfaction with internal or external stressors reaches a certain level, the push towards a change in housing becomes dominant.

Having decided to move, the second phase involves searching for housing alternatives and comparing these alternatives with the actual and desired living conditions. People will only move if the alternative housing satisfies their requirements and reduces their stress. Brown and Moore found that in older age, people were more inclined to move when they perceived large discrepancies between, on the one hand, their current house and neighbourhood characteristics and, on the other, the known alternatives. Age, loneliness, social contact, costs of living and state of repair of home all proved to be important factors relating to both housing and neighbourhood satisfaction. They also found that people's judgements of the appropriateness of the home in the near future had a significant impact on their decision-making process.

Fokkema and Van Wissen (1997) adapted the *stress-threshold model* to explain the moving plans of older people in the Netherlands. In their model, there are three sets of variables:

- background characteristics;
- level of housing and neighbourhood dissatisfaction; and
- moving plans.

Background characteristics are assumed to influence housing and neighbourhood dissatisfaction, which in turn influence moving plans. Fokkema and Van Wissen included:

- demographic factors (e.g., age, number of persons in the household, financial position);
- health and social factors (such as need for care, loneliness, social contact);
- housing factors (including too few rooms, too many storeys, state of repair); and
- neighbourhood factors (e.g., concern about neighbourhood, safety, level of housing and neighbourhood dissatisfaction).

Applying the models

To illustrate a practical application of these 'pull–push' approaches to housing decision-making, some recent work to develop and refine a number of housing option appraisal tools for use by older people will be described (Heywood *et al.*, 1999). Various formats for the tools are being tried: the original self-completion full questionnaire, an adapted version for the internet and a mini-version in the form of a scratch card. The *Housing Options for Older People* (HOOP) tools are designed to provide an additional resource to help people who are considering whether or not to move house. They aim to widen people's thinking about possible problems they may face and identify any extra sources of information they might need.

The tools encourage people to make a subjective assessment of how well their home suits them in terms of nine categories. They are asked to rate the suitability of their home's physical characteristics (location, size and space, comfort and design, condition of property, physical security and safety). Then they are asked to consider how easy it is to live in their home (e.g., to manage in their home or cope with financial costs) and finally, to judge the capacity of their home to promote and maintain their independence, well-being and quality of life. Having done this, they prioritize these categories according to importance. The next stage is to think about, first, whether their home would still suit them if their circumstances changed, secondly, how they view the prospect of moving and, finally, what their options might be.

A critique of rational decision-making approaches

Despite the number and range of decision-making models, it is possible to identify some shared characteristics which allow comment on the level of explanation and choice of inquiry method, the taken-for-granted assumptions, and finally, the notable omissions.

Inquiry method

Different domains of analysis

To make sense of the diversity of approaches used to explain housing decision-making behaviour, it is helpful to classify them according to where they locate their explanations and within which conceptual framework these explanations belong. To aid this task, Sapsford (2003), whilst acknowledging such a classification system is necessarily highly artificial and oversimple, proposes four domains of analysis: *societal* explanations, *group* explanations, *personal/interpersonal* explanations and *intrapersonal* explanations.

Models that make predictions about the moving behaviour of older people in terms of demographic factors such as social class or marital status provide societal explanations. Those that consider the way knowledge of specific older people's needs, such as their reduced mobility levels, could help predict how they will behave, offer group explanations. Neither societal explanations, which examine relations between whole groups or classes in the wider society, nor group explanations that examine the behaviour of subcultures, focus upon particular individuals. Personal or interpersonal explanations treat the person as a whole, as someone who whilst interacting and having relationships with other people, is analytically separate from them. Explanations within this domain tend to be in terms of reasons, or actions rather than causes and behaviours. Finally, intrapersonal accounts look at what goes on inside the person: what they think, believe, feel, decide and do. Here, the focus is upon players rather than the game as a whole. The *Housing Decisions in Old Age* data reported on in this book falls within the latter two domains of analysis.

Phenomenology of individual

Phenomenological research 'seeks to capture as closely as possible the psychological meanings that constitute the phenomenon through investigating and analysing lived examples of the phenomenon within the context of the participants' lives' (Giorgi and Giorgi, 2003, p. 27). Central to this approach, is the recognition of the subjectivity both of human experience and of responses to situations.

Inquiring about the individual experience of making housing decisions by exploring meanings attached to home, perceptions of the neighbourhood and feelings about moving house in later life, reveals the importance of individual differences and of the social, physical, economic, cultural and historical context. Collecting personal stories shows how people with different personalities, intellectual capacities, ages, genders, biographical histories and cultures respond to changing circumstances associated with their own ageing, their housing and their environment.

Assumptions

Are people rational decision-makers?

As we have seen, models such as the subjective expected utility, cost–benefit and pull–push models, assume people are rational decision-makers who assess the probable costs and benefits of alternative courses of action, then act to maximize their highest expected utility (see e.g., Von Winterfeldt and Edwards, 1986). The stress-threshold model assumes less rationality and predicts people will move when the stress experienced is too high.

The question of whether people are rational has occupied philosophers and psychologists for years. Evans and Over (1997) propose that the answer to this question depends on how rationality is defined. Some argue that because humans are the most intelligent species on earth, with remarkable cognitive and communication capacities, rationality is a self-evident human trait. By contrast, most claims of irrationality in psychological experiments are based upon the notion that people are only being rational when they use formal logic or probability theory.

For a number of reasons, problems arise when trying to use these decision-making models to understand everyday decisions. First, whilst many decision-making theories and models intuitively make sense, crucially, they do not account for the fact that individuals define risks and utilities in their own subjective way (Evans and Over, 1997). Thus, older people may continue to live in a house which is too expensive, when the housework has become too onerous, and the public transport inaccessible, because they place a higher value on keeping up the appearance of coping to friends and family. Alternatively, people may say that living where they do gives them such pleasure that it is worth putting up with less money, a dirty house and unkempt garden, and not being able to get out very often.

Secondly, it is clear that humans have the capacity to be rational, that is, to engage in hypothetical thinking to imagine future possible scenarios, using correct deductive reasoning and hypothesis testing to lead to a consequent decision. However, decisions in everyday life are not always made like this. People trade long-term goals (perhaps the peace of mind provided by living in a house or area that can accommodate the needs of older people) for immediate goals (such as enjoyment in living in their current home); normative decision-making theories and models do not distinguish between the two. Evans and Over contend that people rely on past learning rather than a hypothetical analysis of future events.

Considering whether to move house in later life involves assessing the likelihood of being able to continue to live in the present home in the future. One of the main reasons why staying in their home might become difficult is because of a deterioration in physical health. Weinstein however, has found that people tend to have inaccurate perceptions of their own risk and potential health problems, which he terms their 'unrealistic optimism' (Weinstein 1983, 1984). He suggested there were four cognitive factors that contribute to unrealistic optimism:

1. lack of personal experience with the problem;
2. belief that the problem is preventable by individual action;
3. belief that if the problem has not appeared yet, it will not appear in the future; and
4. belief that the problem is infrequent.

These factors suggest that risk perception is not a rational process.

Whilst all of the decision-making models view explanations for moving as multi-causal, they imply a deterministic, linear causality rather than circular (even spiral) causal functions. This over-reliance on linear models of causality leads to a failure to capture the complexity of the shifting interplay between individual and his/her social system. The stress-threshold model is possibly the exception, as it incorporates the intricate relationship between the individual and his or her environment.

Models stemming from a reductionist philosophy

In simplifying complex phenomena and processes, societal and group explanations can be misleading, because they present partial explanations which are divorced from their context. Reason (1994) warns of dangers of overly rational, reductionist approaches which attempt to break down a complex process into parts:

> The tendency to think in terms of parts rather than wholes, things rather than processes: naming the parts of the world creates an illusion of real separate objects; concepts drive a wedge, as it were between experience and understanding. These mental constructs, or paradigms, are immensely robust and self-fulfilling when isolated from experience. (Reason, 1994, p.12)

Housing decision-making models provide useful conceptual frameworks that can inform the thinking of professionals and older people. In our view, however, they should be treated cautiously and regarded only as a starting point. If they are used too rigidly, they necessarily lead to a restricted outlook that does not accurately capture the older person's unique understanding of their situation and, consequently, risk being seen as unappealing and therefore rejected. There needs to be scope for people to volunteer and freely express their own view of their situation, before being steered in a certain direction, which may not fit them.

The gaps in rational approaches

They do not accommodate direct effect of impulse or intuition

Intuitive thinking is used in everyday life. People make some decisions impulsively, in that they are not pre-meditated: they rely on gut feelings, reactions, hunches or what they might term common sense. Intuitive people may have a heightened sensitivity to clues, be extremely attentive to detail and may see meanings in things which others overlook, all of which may occur consciously or subliminally (Claxton, 2001). For these reasons, individuals may try to articulate what was going on, but perhaps find they cannot. Of course, people differ in their intuitive ability and the extent to which they trust their own intuitions. Jung used his clinical experience to identify 'intuiters' as one of his four personality types. 'Intuiters' prefer to take information via a sixth sense of what might be; they like to see the big picture and can tolerate change, uncertainty and confusion. They can elicit information that is beyond that available from the senses, such as being able to read between the lines and look beyond the surface. These skills involve perceiving relationships, imagining consequences and anticipating possibilities.

Westcott (1968) working from experimental evidence labelled some participants 'intuitives', in that they were willing to answer on the basis of little information rather than being cautious and waiting until they had more information. These people did consistently well at solving problems with only the information they had asked for (Cited in Claxton, 2001). Westcott found that intuitive people tended to be introverted, self-sufficient, trusted their own judgement, enjoyed taking risks, were able to make-up their own mind and resisted being controlled by others.

Intuition has often been viewed with suspicion and seen as rather mysterious and the very opposite of reason. We do not intend to be anti-rational or anti-intellectual, but want to highlight the contribution of subtler, and complementary, ways of understanding decisions. As Atkinson and Claxton have said:

> Intuition can provide a holistic way of knowing – it appears to be unconscious insight but it is not therefore without basis. Rather, its basis is the whole of what has been known but which cannot, by nature of its size and complexity, be held in consciousness. Reason, by contrast, is concerned with conscious analysis of knowledge which confines it to the level of detail. (Atkinson and Claxton, 2001, p. 5).

They do not accommodate emotion

It is our belief that emotional aspects of making housing decisions should not be overlooked. Intuition and emotions are closely linked. When people cannot decide between competing options they often make decisions based on what their emotional reactions tell them. Thagard (2001) suggested that people are likely to be motivated into action when their decisions are based on emotional intuitions. He proposes another advantage:

> Basing your decisions on emotions helps to ensure that the decisions take into account what you really care about. If you are pleased and excited about a possible action, that is a good sign that the action promises to accomplish the goals that are genuinely important to you. (Thagard, 2001, p. 2).

In a rational decision-making approach people will systematically identify the options, making a list of acceptable criteria and then judging whether each option satisfies the criteria. The difficulty is that people may then react to the winning choice with, 'But I don't want that'. Thagard puts forward his answer:

> Your emotional reaction need not be crazy, because it may be that the numerical weights that you put on your criteria do not reflect what you really care about. Moreover, your estimates about the extent to which different actions accomplish your goals may be very subjective and fluid, so that your unconscious estimation is at least as good as your conscious one. (Thagard, 2001, p. 4)

They do not address the issue of joint decision-making

Housing decisions are rarely taken alone and the rational models do not acknowledge and account for influence of other people, whether that is from a spouse, children or other close family members. Some people actively seek out the opinions of others, wanting their expertise, experience or inside knowledge. In addition, the views of other people may matter a great deal to them and need to be taken into account. There are some who will experience attempts at persuasion as an uncomfortable pressure that they find difficult to ignore, even though they may question whether the person offering advice is pursuing their own interests. Such perspectives could be incorporated into a rational model, but to date they seem to have been neglected.

Cognitive capacities of older people. Family systems theory suggests that two aspects of family history may become relevant in later life: 'how previous generations dealt with ageing-related changes and how this particular family constellation adapted to previous life transitions' (Smyer and Qualls, 1999, p. 117).

The discussion of the general features of ageing which follows must be placed within the framework that individuals all age differently. Physiologically, ageing manifests itself as a general slowing down of functions combined with an increasing vulnerability to damage and disease, although not all biological functions will be affected to the same degree. Physical strength is one of the functions that declines quite dramatically, although the impact on the individual will be dependent on their prior levels of fitness or 'reserve capacity'. It is similar with changes to the brain, where a generalized effect of ageing is cell loss in key regions and widespread shrinkage of the brain tissue, but compensatory strategies and a prior high level of functioning may mitigate the effects.

However, the brain is not infinitely flexible and for some losses there appear to be no compensation mechanisms. Even a slight decline in a single function may have a significant impact on a person's ability to undertake complex decision-making tasks. This is likely to lead to an increasing reliance on stereotypes, heuristics and schemas in decision-making, even in situations where detailed analytical processing would be more appropriate.

There is evidence also that older people may become more easily distracted from the task in hand as there is a decline in the brain function which selects and screens out irrelevant stimuli (e.g., studies cited by Park and Schwarz, 2000, p. 107). In addition that as the day progresses, information from activities that have taken place increasingly interferes with the processing of new tasks (Park and Schwarz, pp. 151–71). The 'interference effect', in itself slight, creates the phenomenon of 'morningness' among older adults who may be less well equipped to carry out demanding cognitive tasks later in the day.

Mood and motivation also play a major role in decision-making and here evidence from epidemiology suggests further challenges for a significant minority of older people. Levels of depression amongst older people are higher than for the population as a whole and may lead to a lack of motivation and social withdrawal (Smyer and Qualls, 1999, pp. 7–14). Mood does affect decision-making capacity, perhaps leading people to rely on their own knowledge rather than others', or to put off making a decision all together.

Consequences for decision-making

From a consumer perspective, decision-making seems straightforward: people need wide-ranging and high quality information; from this they will be able to process the data that they receive and emerge with the best solution for them. As we have argued in this chapter, decisions rarely fit this simple outline. The giving of more information adds to the complexity and creates further tensions as to which is the most important criterion.

Thinking about experiences in making other consumer decisions can help illustrate this point. Attempting to be a model consumer when buying kitchen equipment such as a cooker is complicated: having taken advice from *Which* and other consumer guides, discovered that the model referred to in the report differs from a current model on display, considered which of the criterion used matters most, there remains the searching for the best deal in terms of price, delivery and aftercare.

Going to a local shop and taking limited advice often seems the easiest option. It is little wonder that, when deciding where to live, people may close down the range of options or criteria. It is important to recognize that many people do not make their housing decisions in the way that the consumer ethic would predict.

Heywood and colleagues set out the factors that affect decisions to move or not to move, heading the list with 'the reticence or inability to expend the physical and mental energy required to undertake a move' (Heywood *et al.*, 2002, p. 81). Such reticence or inability may be the consequence of depression, as mentioned earlier, but equally may be because the demands inherent in being a consumer are too great.

Alternative approaches to understanding housing decision-making

Life-course, longitudinal approaches and housing biographies

Clapham and colleagues (1993) use the concepts of 'housing career' and 'housing history' has countered the rather static approach taken in previous research on modelling housing decisions. Their focus has been more dynamic and has shifted the focus towards looking at how an individual's previous housing pathways, particularly their life events, socio-economic circumstances and use of coping strategies, have led to their current housing situation. Payne and Payne (1977) conducted a series of cross-sectional surveys to examine the progress of households and their movement between housing tenures. Households were selected from a range of initially different housing situations and then tracked from the birth of their children but, unfortunately, this was only for a limited time period. Forrest and Kemeny (1984) argued that the housing career concept did not match the experience of those who reacted to crises and unplanned changes in circumstances, as opposed to purposefully and strategically planning their housing moves. Forrest and Kemeny's solution was to combine the concepts of housing career, constraint and adaptation via coping strategies. Forrest later abandoned the idea of housing career in favour of housing history, showing the importance of employment histories in determining individual housing pathways (Forrest and Kemeny, 1984 and Forrest and Murie, 1985, 1991 cited in Clapham, 1993).

The potential for longitudinal research, based on biographical interviews, as a means of understanding housing choices in later life has been advocated by Clapham and colleagues (1993). They suggest that making the links between various elements of people's lives is crucial in understanding how people arrive at their present housing situation. As longitudinal research is difficult to fund, cohort study research designs could provide a promising alternative, allowing groups of people, usually of different chronological ages, to be compared.

Thinking holistically

When housing decisions are viewed as a whole, rather than analytically, new possibilities emerge. Holism has widespread appeal within many diverse fields, and has perhaps become rather tired and overworked, with an endless list of professionals claiming to offer 'holistic approaches', 'holistic medicine' and 'holistic assessments'. Nonetheless, conceptually it remains a useful way of capturing the essence of our argument. We use 'holism' to mean *everything to do with something* that goes towards its final make-up, namely, the eventual housing decision. From an individual perspective, this allows the whole person to be considered and allows the inclusion of both real phenomena *and* intangible phenomenon, such as perceptions, attitudes, beliefs, ideas, emotions, personal meanings and the influence of others. It embraces different ways of knowing, integrating rational, reflective and intuitive styles of thinking and responding. Finally, it also recognizes the wider context and social systems in which housing decisions are made and the complementary existence of societal, group, as well as intrapersonal, personal and interpersonal explanations.

In the next chapters, we illustrate and expand upon the themes highlighted so far drawing on data from in-depth, individual interviews and housing biographies. By asking participants to share their experiences of making housing decisions, we have begun to see how and why people plan or react to circumstances: some think ahead about strategies for managing their later years, others await events. The influence of different personalities, perceptions, attitudes and beliefs on the shaping of housing histories is also evident. Expanding the remit to include their views on what constrained their housing choices and what happened when they nearly moved, but decided against moving, has also proved fruitful.

In the next two chapters we argue that the complexity inherent in decisions about housing, and in their evaluation, can only be appreciated when people's individual emotional lives are understood. We contend that many attempts to explain and even predict housing choices in later life are problematic because they do not pay sufficient attention to older people's emotional lives. In making housing decisions, people attempt to identify and weigh up the pros and cons: whether they can cope with the financial costs associated with owning their own house on a reduced income, or whether they can cope physically with the home and garden maintenance as they grow older. Undoubtedly, these factors are important, but they are only part of the explanation of the making of housing decisions.

Emotionally, people may be motivated to make decisions that initiate change or maintain the status quo, and emotions may also affect the final decision from a range of options. Perhaps more important is the relationship between emotion and the perception and assessment of one's situation. There is, for example, a large body of psychological literature which examines the effects of emotions on cognitive processes. Furthermore, radical psychologists and some social constructionists argue that the emotions do not merely impact upon social lives and understandings of reality, but are constitutive features of these relationships, and indeed constitute reality itself (Gergen, 1994).

With regard to housing, the emotional context of people's housing decisions includes fears, feelings of safety and insecurity, beliefs, aspirations and interpretations of the past, present and future. Many people have deep attachments to their homes and neighbourhoods. Their sense of identity is bound up with the feelings they have for places and people. Some older people express a yearning to return to places they associate with childhood memories or to what they perceive as their roots.

For anyone, moving house means acknowledging and dealing with loss of significant people, places, memories and possessions. In later life moves may occur at a time when people are managing other difficulties, perhaps from illness, disability or loss of relatives and friends. For some older people, moving into specialist housing accommodation may mean changing the way they, and others, see themselves. They may put at risk their perception of themselves as an active, independent person who is in control of their life and lives in their own home. In return, for some people, their new self is someone who is a less active, less independent person, living in a place which feels as if it is someone else's home (Clough, 1998).

To illustrate the salient role of emotions in housing decisions, in the next two chapters we shall address particular emotional themes that have had a noticeable impact on the decision-making processes of research respondents.

5
Attachments to Home

The last chapter focused on the process of decision-making, the ways in which people decide what to do. In the next two chapters we look at some of the less tangible reasons why decisions are made. First, we consider what is being constructed as people live with the decisions they have made about their house. This is followed in Chapter 6 with an examination of another part of the backcloth: people's worries about their lives and how these influence housing decisions.

The idea of 'home' and what it means to people has been the focus of considerable literature in the social sciences. Despres (1991) points out that the meaning of home exists on a number of spatial scales, such as neighbourhood, region or country. Our focus is primarily on the meaning of the home itself, although people attach to that building feelings about the place in which it is located. In particular we look at how older people feel about their home and their affinity for the places in which they live: a positive sense of belonging and emotional attachment with spaces they consider their home which in many cases they have built up over a considerable length of time.

Despres's (1991) categorization of the meaning of home amongst the general population is widely cited, but remains a useful overview:

- as security and control
- as a refuge from the outside world
- as a permanence and continuity
- as a reflection of one's ideas and values
- as something 'acted upon' by the person
- an indication of personal status
- as a place to own
- as relationships with family and friends

- as a centre of activities (a base of life)
- as material structure.

With regard to the meaning of home specifically to older people, Pastalan and Barnes (1999) identify six key points:

- home as a legacy and a repository of history;
- home as a reflection of the occupant in terms of social status (and its potential decline) or a life project;
- use of home for relationships with others and the outside world, either as a social centre (bringing in) or as a refuge (hiding from);
- ability of home to provide privacy;
- home as a space where daily activities, rituals and identity are enacted, and the familiarity which goes with this;
- ability of home to provide a sense of control and confidence for older people.

Such analyses suggest a degree of deliberation: in reality, most people get on with the daily routine of living and of organizing their house to suit the way they want to live. A few may be aware of working to create security or a refuge; more, on reflection, would be able to find phrases to describe what they have been doing. In this chapter we focus on the ways people use their house to create their place in the world.

Many of the studies of meaning of home in later life have focused on owner-occupiers and their relationships to the houses they own, particularly in terms of security, independence and legacy. Askham and colleagues (1999), in their study of owner-occupiers, found three important factors were cited: independence, finance, and sense of identity. Independence was found to be double-edged in that in addition to enabling independence, older occupiers often felt restricted in terms of maintenance, financial duties and freedom of movement. Homes were seen as providing financial assets and security. They were intensely bound up with a sense of identity among older people. Home owners were proud of having 'achieved something' and felt more secure as a result. However, with increasing disability, divorce or widowhood, home and the responsibilities tied to them were perceived as more burdensome.

Dupuis and Thorns (1996) conducted a similar study in New Zealand. They found that meanings of home for older home owners could be placed in three broad themes:

- An association of home ownership with the quest for *security* (financial and otherwise).
- *Family continuity*: 'A home therefore was understood as a physical building, a house, transformed over time into a home by the presence and practices of the occupants, the family' (Dupuis and Thorns, 1996, p. 496). Thus there is an association of home with memories and nostalgia.
- The combination of the two points above in terms of *inheritance*: the home is something to pass on to children.

Gurney and Means (1993) argue that the study of home in later life should adopt a more experiential perspective. Documenting the rise of home ownership in Britain in the 1980s, they propose the development of a new model, one which recognizes that meanings of home are specific to the experiences of certain moments in time and are constantly changing. They also note that meaning partly emerges from the intertwining of people's life histories and their houses. They suggest that houses often embody the substantial efforts of 'sweat equity' (and in that sense can be seen as lifetime projects), and that houses of long tenure can act as anchors for the mental maps of older people. In this sense, 'home' in older age can be seen to have three levels of meaning:

- the *cultural level*: that is the emphasis of 'home' and 'ownership' in our culture;
- the *intermediate level*: the everyday level of practical life; housing histories or careers, length of tenure, housing decisions; the 'real life' situation versus a person's aspirations and expectations; and
- the *personal level*: the symbolic and personal experience of 'home'; important events in life history.

Studies like these have concentrated on what Despres (1991) would call territorial and psychological interpretations: a home is something one possesses, and as a possession, is secure and an expression of oneself, one's life history, and one's accomplishments and position in society. Dupuis and Thorns (1996) have pointed out that such a strong attachment to home ownership is specific to the current generation of older people. They argue that later generations of homeowners are not as likely to have as profound an attachment to 'home' as a fundamental part of their lives due to changes in family structures, and a more mobile society.

People are moving more often, perhaps to change jobs, and consequently moving away from families. The influence of socio-economic factors has been undervalued in current literature. For example, it is widely noted that those with higher incomes or more formal education are much more likely to move regions or even climates after retirement, or to purchase second homes.

Some of the accounts of housing from our study allow greater searching at what makes people feel comfortable or uncomfortable in where they live, and how this affects their overall well-being. Such an intimate understanding of what is important about 'home' would be labelled by Despres (1991) as phenomenological – how home is 'done', for example, through actions and rituals, in a familiar environment. Willcocks and colleagues (1987), in their study of residential homes, argue that familiarity and privacy are particularly important for older people in relating to home:

> With the privacy of home, an older person can control, and often conceal, declining capabilities in the management of daily living. The familiarity of the setting permits what Rowles calls a sense of 'physical insidedness' where familiarity, at least at the conscious level, can compensate for the progressive sensory loss that is likely to accompany old age. (Willcocks *et al.*, 1987, p. 7)

Similarly, Pastalan and Barnes (1999), who also see the home as a locus of order and control in a world of uncertainty, stress the importance for older people in residential care of being able to keep personal rituals. They argue these are important life-centring activities which generate a feeling of home. Loss of these, they maintain, leads to a diminution of the self.

'Home' can refer to the *structure* (the physical building), including the *objects* (the possessions within a home, symbolizing a life history), and the *environment* (the 'place' where the home is located and the social environment of friends). Attachments to home that we found evident in our research can be grouped in the following ways: home as privacy and sense of control, as a collection of possessions, as an embodied or an emplaced relationship, as a centre for daily activities, and as a social network.

The remainder of this chapter will discuss these different meanings in turn. It explores the way housing can be understood at both the *intermediate level* of meaning (i.e., is how it is experienced at an everyday level of practical life) as well as the *personal level* of meaning, the symbolic and personal experience of 'home', together with the importance of events in life history.

Home as privacy and sense of control

Individuals describe home as being a familiar, comforting place where they have some control over when, or indeed whether, the space is public or private. They describe how it can function both as a social space for entertaining and inviting in family and friends, and also as a place of refuge from other people and the outside world.

> To me this house is where I am comfortable, feel generally safe. I know that's a slightly different aspect but, no, it's where I am happy, it's where I've brought my family up, where mum and dad used to come. I mean they loved it – they couldn't understand us wanting to move from where we were. But when they came, oh they used to love it here. To me, it's, well, it's roots almost, isn't it? We've lived here longer than we did when we were in South London. I mean I was born there, what 1930, and we've been here since '64. No, to me this is home. I mean ... even our two boys they talk about going home but I can't think of any other, to me it's us, it's here, it's our family home. So it's a comfort sort of place. (Mrs Oliver, 71 years, married, 3-bedroomed house, London)
>
> Oh yes, yes. And the other lovely thing is you can come and go as you like. If you want your meal in your room, you can have it; if you have friends, you can entertain friends if you want. You can go out, you know. I mean, you have a door key and you can come and go. It's like being in a very nice little hotel, you know. (Mrs Waterford, 81 years, widow, residential home, London)

For Mr Read, what is treasured is the ability of his home to provide a retreat that reminds him of his personal history:

> Our home is, an old phrase, is our castle. Our home is our place of refuge, a place of comfort, a place of love. Well, a place of comfort. We want our things around us. They all mean something. Our home is very special to us. It's not just bricks and mortar. It's the love that goes into a house to build a home. Again, that is something that researchers don't always appreciate. It's what you put into it and it's your roots. (Mr Read, 67, Single (living with mother), owner 5-bedroomed home, Lancashire)

For those who were currently living in communal residential settings or had done in the past, having control over whom they lived with and indeed, whom they would prefer not to live with, was an issue that set apart communal residential settings from more private forms.

> But in a residential home you can't do anything about it, can you? If a person is in need of coming into a residential home and there's room for them, they've got to be admitted, haven't they? And you can't turn round and say well in my opinion, 'Oh no you can't have her because ... ', if they're ill or something like that, that's a different matter. But temperaments, you can't do anything about that can you? (Mrs Beeforth, 83 years, divorced, sheltered housing, Lancashire)

A member of one of the advisory panels, described how, having grown up in a children's home, she now had a mistrust of such establishments and dreaded ever having to enter a residential home.

> Your home is your sanctuary, where you can have people that you like, you turn away people you don't want. It should be your own private place to live, and be emotionally yours as well. (Leicester older people's advisory panel)

This wariness of the intrusion of others was echoed by Miss Davidson who, having moved into sheltered accommodation at the unusually young age of 55 years, found the pressure to conform to the wishes and expectations of other sheltered housing residents distressing, feeling as if she was under surveillance. She speculated that residential homes would accentuate this type of problem:

> I got very depressed the first six months because, the neighbours (were) always wanting you to go in for coffee. There's a couple of them, not so much as a mouse gets past the window, and that can get very wearing, to know that you are under scrutiny every time you come in and out. It would really be rather nice if we had another entrance. You're timed if you go into someone else's house, and it can really, really get you down. They've got used to it now. I've just had to be very firm and say I'm here if you want anything, but I really don't want to get involved in and out of everybody's houses every day. I just couldn't be doing with that.

Later she states:

> You see, if you've your own place, if you want to play a little bit of classical music like I do, I know you can't play it loud even if you've got your own place, but you are, you sort of think that you're a lodger in a big house and everybody's living together. And you're not, you're not a person, you're part of this big house. (Miss Davidson, 60 years, single, living in council owned, sheltered housing 2-bedroomed bungalow, Cumbria)

Residents from nursing and residential homes spoke of feeling uncomfortable after finding they had little or nothing in common with other residents, or of not being able to communicate effectively; they resorted to withdrawing to their own rooms.

> I'll be honest with you I do not go into the lounge a lot because, it sounds as if I'm being strange and I wouldn't hurt anyone, but you can't have a decent conversation with them. (Mrs Cadman, 71 years, nursing home, Lancashire)

> I'll never fit in properly with some of the people here because they don't hear me. They're not on a wavelength of any kind. If you say to them, 'Good Morning', they'll say, 'What did you say?' So you see, I've stopped saying 'Good Morning' to a lot of them. So I go down for my breakfast and pretty quickly, I have it quicker than anyone else, and I come back here or I go out with my daughter or with some friends. I really get on better with the manager here, and the top person in the office upstairs, I get on better with them, because we can talk. But of course, they are busy half the time. (Mrs Ashley, 90 years, widowed, residential home, London)

In these cases both Mrs Cadman and Mrs Ashley have retreated out of the communal aspects of their living situations because they resent the interactions they feel they are forced to make when using more communal spaces. In that sense, they feel that they have a lack of control in communal living compared to living privately when they could associate with others on their own terms.

Home is viewed as a place to exert control and autonomy, a setting in which people can live their lives as they wish. One motivation for many to stay in their own private homes is that they feel they are able to do as they wish in a way that they would not be able to in more communal settings.

Home as a collection of possessions

Places, and particularly certain objects within those places, can remind people of their past and themselves. Possessions can, in a sense, 'narrate' people's life to them and reluctance to part with things is an important factor for some in making housing decisions. Hormuth (1990) proposes two ways in which objects could be said to help maintain their owner's identity. The first highlights the role of objects in the construction of the life history and historical continuity of individuals (see Dittmar, 1992, p. 206; Hormuth, 1990; Silver, 1996).

The second involves a relationship in which objects establish an 'ecology' from which a sense of identity is constructed and maintained. Objects are seen not merely to provide links with the past in a narrative fashion, but to create an environment referred to as an 'ecology of the self'. Like an ecology in nature which over time achieves a balanced coexistence among plants and animals, a balance is established between the conception of one's self, and the surrounding objects and environment (Hormuth, 1990; Rochberg-Halton, 1984).

These two themes, the role of objects in the construction of identity and their part in the establishment of an ecology, are both to be found in our research. They are discussed in turn.

In the narrative dimension, objects are symbols and representations of social experiences, of the past in general and in particular. In life, the roles of individuals change often and rapidly, from a scale of hours in a day (e.g. from worker to parent to spouse) to one of decades (changing from child to student to worker to retired person). This is particularly true of older people, who face the transition from working to retired life, and perhaps from being one of a couple to living without a partner.

Within this seemingly endless process of role transformation, objects can help maintain our sense of who we are by providing coherent documentation of prior stages and events in our lives. This is done through the use of objects as *anchors*, termed such because they are associated exclusively with prior stages of life. Anchors are akin to what have also been called *transitional objects*. Silver (1996), for example, argued that anchors were often crucial to the identity and history of the university students he studied: objects brought from home could be reinterpreted to be consistent with new identities as students, but also provide documentation of previous events in their home lives.

More eloquently, de Certeau (1984) wrote that objects have a hollowness in which the past sleeps. Objects are the words in which we recite the diary of our own lives. One of the research participants wrote:

> I am standing in my untidy workshop, using hand tools, many of which were my father's. I can feel him behind me, correcting the wrong use of things like files. The vice and workbench are survivors from the earliest days of our marriage. I repair many household items. I make devices to cope with contingencies, for example, as a means of enabling a disabled dog to get in and out of the back of the estate car safely. Finding solutions to problems, often by improvisation, gives me a connection with my working life as a method study engineer and development technologist in the textile industry. (Mr Blake, 67 years, married, 5-bedroomed house, Lancashire)

Mr Blake described how he uses his workshop and his tools to establish continuity not only between his working life and his past in general, but also between his past and his current retired life. Through his interaction with these objects he is able to maintain his sense of himself as a practical, problem-solving person.

The crucial point is that objects used in such a way provide evidence or documentation of an individual's ongoing life story. In another example, Silver notes that older people often put a great deal of investment and effort into communicating their own life stories through their collections of objects, sometimes taking great care to explain the significant events or persons behind each one to those that will inherit them. In the extract below from our research, Mr Selby explains the significance of his collection of Indian artefacts:

> It's a comfort factor, yes. And any sort of knick-knacks and things I have around here are of sentimental value … tiny little Indian things which people gave me when I was working in India. They have no value, so that's why I don't worry too much about burglars, but one is surrounded by sentimental objects. (Mr Selby, 74 years, single, 2-bedroomed flat, London)

By surrounding ourselves with the objects of our past, we try to master time and the march of our own lives. The past becomes a series of moments which we can recall for ourselves and others, and these moments become embedded within our collections of sentimental objects:

> We've always lived in larger houses, had a four or five bedroom house and all the rooms (that are) needed, a dining room and study, things like that. And if you've always lived in them, it's very difficult to go into a smaller house because over the years you collect things. I will show you in the other room afterwards, I'll show you what I mean by 'collect things'. So in a smaller house, what possessions would we have to part with that have a meaning to us? Everything in this room, it might be meaningless probably to you, but to us, it all associates with something, with our past life. And you feel that if you get rid of that you are getting rid of part of your life, as strange as it may seem. But when you are old you will remember those words. Look in this room – this is but one room. How could we move all this stuff now? Everything in here, those three racehorses, these are what they call the three kings – that's Red Rum, Arkle and Desert Orchid. They have meanings. To my parents all these things have meaning. Now if we decided to move to a smaller house, we would have to get rid of these wouldn't we? And then you're getting rid of part of my life with my father. (Mr Read, 67 years, single, living with mother, owner 5-bedroomed home, Lancashire)

Mr Read and his mother plan never to move. Their house is crowded with physical expressions of memories. So they endure the costs and effort of keeping up a five-bedroomed house, even though they have to struggle both financially and physically. They choose to stay put, adapt, and have a lower standard of living rather than part with the possessions which embody so much of their past lives and memories.

A notion of self or identity is re-enforced by the familiar objects with which one is surrounded. A person invests part of their identity in their surroundings so that these surroundings can in a sense 'speak back' that person's identity. Moving house has the potential to upset this balance:

> The self concept exists in interdependence with its ecology of others, objects and environments. As long as the ecology of the self is stable, the self concept will be stable and strive toward maintenance. Self concept change, on the other hand, results from an imbalance in the ecology of the self that leads toward restabilization under different ecological conditions. (Hormuth, 1990, p. 3)

Rochberg-Halton takes a similar approach and notes that household objects:

> ... form a part of a gestalt for the people who live with them – a gestalt that both communicates a sense of home and differentiates

> the types of activities that might be more appropriate for one part of the home than another. Hence the meanings of the things one values are not limited just to the individual object itself, but also include the spatial context in which the object is placed, forming a domain of personal territoriality. (Rochberg-Halton, 1984, p. 352)

Thus, homes are not only collections of objects that relate to specific past experiences, but provide an environment which speaks to the occupant. Their continued use provides reassurance and continuity. Some things have always been present, perhaps a comfortable chair or a bookcase. Interacting with these things is akin to having a conversation. A person extends themselves into these objects, and at the same time, these objects become a part of the self. As a result, many people become attached to the things they have acquired over the years, seeing them as integral to their sense of identity. Mr Borwick and Mrs Murphy provide examples:

> As I say, our son would have done this or would have wanted us to move to a bungalow. ... And he can see it in very wrapped up terms, you know: 'You go here, you move to a bungalow, you get rid of your excess furniture, you get rid of this, that, and the other'. We see these things as ... part of us. If I've got to go and buy a smaller armchair because the one I'm used to won't fit into a small bungalow, you know, I don't want to do this. I want to hang on to being me as long as I can. (Mr Borwick, married and interviewed with wife, living in own home, Lancashire)
>
> I think it's the things that I have got that have been with me all my life like even just silly things like that book case. It was the first thing I bought for my first house because my books were all on the floor. And although it's not of any special value, it goes everywhere with me and you feel these are all my things. I'm at home with them. But ... wherever I've been, they've been with me. I've moved house quite a lot of times. I've lived in a flat, in a terraced house, in a semi, in a detached house and so you get attached. (Mrs Murphy, 60 years, married, living in bungalow, Cumbria)

Ultimately though, there are those who have to face giving up possessions. Some have found this difficult, but many have been able to reach satisfactory compromises. The interviewee below describes how she came to terms with having to part with her furniture:

> I had a large flat in London, two bedroom, large ground floor flat. ... And I had ... loads of furniture, two big dressers for instance.

> So the decision had to be: 'What did I really need in this tiny little cottage and what would I do with the rest?'. And so I decided what I really needed. And the basics for me are I need a bed and a good comfortable chair. I've seen that with my granny who lived till a hundred. The one thing she needed was a good comfortable chair. I need a television, because I think that, as you get older, that becomes your companion. For me, I'm sort of half in the modern world as well, so I need my computer. But I didn't need any more (*furniture*) so I told my children, if they wanted any of it they could have it, otherwise I would try and sell it. But in fact they all wanted it, so it was all divided amongst my children. So when I visit my children, I visit my furniture. (Mrs Raynes, 63 years, widowed, 1-bedroomed bungalow, retirement village, London)

Others appear to adjust by reassessing the importance of material possessions in their lives, now that they are older. Using the analogy of life as a book, Mrs Panton reflects upon the different chapters of her life and why she now regards her present chapter as the best one so far.

> *I*: So do you think a residential place like this is something that some people are suited to and some people aren't?
> *Mrs Panton*: Yes, I think so. Yes, I do think that. Some people will hang on to their own houses and partly because they are bound by their possessions, they don't want to let go. But we've all got to let go. We don't bring anything into the world, we don't take anything out of it. So really possessions are only there for our convenience, not to bind us, which they do with some people.
> *I*: So you didn't have a problem when you moved here from a house in which I assume you had a lot more things?
> *Mrs Panton*: No. I just gave them away. I stood in the kitchen with my grandchildren saying, 'Anybody want this, anybody want that?' and it was lovely.
> *I*: So it wasn't hard?
> *Mrs Panton*: It wasn't traumatic, no, not a bit. I just thought, it's the next chapter of my life and it's been a wonderful chapter, it's been the best chapter so far anyway. I'll tell you that. It's been wonderful. I am constantly thinking how fortunate I am. I never imagined I'd end up in a place with views like this (*over Cumbrian Fells*). I grew up in Liverpool and life wasn't happy but no, it doesn't matter what the cost is. It's quite good in a way because you think, 'Oh, think of that now, and here I am here'. It's wonderful. It's no thanks to me, it's

> like, going with the flow. I think life opens up if you let it. If you're determined to keep your hand on the tiller and say I will go this way. I've done that in the past and made a lot of mistakes.
> *I*: To let circumstances make the decision for you?
> *Mrs Panton*: Much better.
> *I*: But it drives other people crazy.
> *Mrs Panton*: You begin to accept it when you get to my age. Say everything's worked for you. (Mrs Panton, 83-year-old widow, residential home, Cumbria)

Mrs Panton's response was to end up feeling free of her possessions, and possessions in general. Her acceptance that 'we've all got to let go' is contrasted with her knowledge that many people get bound up in their possessions. For her, this has been a positive experience. She also suggests that her laid-back personality, 'going with the flow' as she puts it, helped her adjust to her new lifestyle.

Thus, for many older people the idea of home is bound up with the objects that their home contains, which they come to see as essential to their sense of self and identity. Objects help people to tell the story of their lives and provide a map to locate them in their place in their world.

Home as embodiment and emplacement

Homes physically reflect their occupiers not only in terms of the objects contained within them, but also in how they are used, adapted and altered to fit the people who dwell within them. They are a physical reflection of people's selves, values and bodies. Handrails become smoothed over time by grabbing hands, steps become worn, chairs and beds moulded to the shapes of their occupiers. In this sense, people can become attached physically as well as emotionally to places, especially their homes. Indeed, it becomes difficult to separate the two when the physical aspect of 'dwelling' and the emotional aspect of 'attachment' become as intertwined as will be shown in some of the testimonies below.

In *Distinction* (1984), Pierre Bourdieu argued that bodies are at the same time representations and represented. What he meant by this is that our working lives, class, tastes, preferences, dispositions and values (what he called the *habitus*) are *embodied* in human beings in terms of body posture, deportment, style and gait, and the overall way individuals carry themselves (Williams, 1995).

The repetition of acts involved with living our lives, working, playing, eating and so on, becomes 'written' on our bodies in what Bourdieu calls the *bodily hexis* (1984; 1977). The body is a 'tool for the job'. For example, a bricklayer develops a particular kind of body, with strong, muscular hands and a heavy back.

> Bodily hexis speaks directly to the motor function, in the form of a pattern of postures that is both individual and systematic, because linked to a whole system of techniques involving the body and tools, and charged with a host of social meanings and values. (Bourdieu, 1977, p. 87)

In this sense, a body is a bridge between its use to do a job (e.g., bricklaying), and the continued preference towards using a body in that particular way. This allows and enforces the status quo. The body is caught up in a 'framework' in which its 'design' determines its use. Thus, bricklayers are likely to keep being bricklayers because they are built for it: their bodies have been 'customized' to that job.

In the same way that people's bodies reflect their lives, so too do places. *Emplacement* (Laws, 1995) describes the relationships between the body and the surrounding environment, the way a place reflects or is made special for the person:

> Landscape and body are indivisible, architecture embodies social relations, and identities are emplaced in landscapes. (p. 274)

Emplacement can be seen as the body's ability to produce and mould a place so that it plays a part in maintaining identity. Places (be they houses, offices or landscapes) put identities into place; in so doing, they shape them, encouraging certain activities and discouraging others. At the same time they are also shaped by the people who inhabit those spaces. Places, like bodies, are shaped by class, values, tastes and the like, and then in turn shape those preferences.

Discussions of emplacement have thus far been limited to examinations of larger 'landscapes'. Laws (1995), for example, studied Sun City retirement communities, where the image of the 'independent, active' aged person is embodied in the manufactured landscape of single-family dwellings and golf courses, signifying financial and physical independence, activity and leisure. The same approach could be used to examine any special housing for older people: the 'landscape' is designed to accommodate certain conceptions of the ageing body through things like sit-down baths, wide doorways, call buttons and security systems.

These concepts could equally be applied to people's homes. People who have 'worked' on their home – altered it, decorated it and put 'themselves' into it – feel emplaced within it. Its very structure symbolizes their values, goals and dreams, becoming a very powerful attachment emotionally and physically. To leave behind the familiarity, effort and adaptations is almost like leaving oneself.

In the following extract, Mr Mansfield, having spent nearly a third of his life faithfully and creatively restoring the building to its original state, sees his home as an extension of who he is. To him, his home is partially represented by the physical structure and partially, by his possessions kept within.

> *I*: In your own words, what is it that's special, what does your home mean to you?
>
> *Mr Mansfield*: I bought the place 24 to 25 years ago. It was tumble down. I have always been, through my career and through my work (as an artist), a creative person. Now I've created this place from virtually a tumbledown place, restored it, so I'm living among my paintings and sketches in a place that I've also created, based upon what it was like some centuries ago. I am living in a part of the history of this country, and I want to pass that bit of history on in good condition to future generations. So I don't know whether I can answer your question any better than that. I feel that since I have restored the place, I can remember what it was like, I can now go around the place in comfort and thoroughly enjoy what I've done to it, or what I've restored. Up at the back of the garden there is the old privy up there you know.
>
> But I have a feeling for the place too. I've never experienced any ghosts here or anything like that but I am sure in years to come someone will experience my ghost inside this place. I hope it's a pleasant ghost. If you took me away from this place, well I think this is who I am I suppose. I can't explain it any better than that. It's my raison dêtre, this is the reason for my life, part of the reason, so remove me from here, you remove me from my raison dêtre, and that I think is it.
>
> *I*: So I guess it sounds like it would be pretty troubling for you?
>
> *Mr Mansfield*: Oh it would, it would. I know that if it had to come, I'm man enough to know – through my life I've had to make a lot of decisions – it's a decision I'm going to have to make, or I hope I'll never have to make but I may have to make.

> *I*: You are prepared to make it if you have to?
> *Mr Mansfield*: Yes, that's right, one way or the other. I would ensure that I can stay here by whatever means – either getting some poor soul to come in and do my driving or – maybe not live here – but who is prepared to come up for a sum of money and do that sort of thing for me. There are options. I don't know which one. It will depend upon my circumstances at the time, what degree of incapacity I have, God forbid. (Mr Mansfield, divorced, owner-occupier, house, Cumbria)

While Mr Mansfield may be a somewhat spectacular case, other people also spoke about the home's physical interior having personal significance. In the example below, Mrs Farris explains how her husband's extensive do-it-yourself work on the house reminds her of him and their relationship.

> Well, as well as all the other reasons for thinking of this house as my home, it suddenly dawned on me when I was standing in the kitchen, ... all the tiles my husband had put on the walls, the cupboards that he put up. And then I thought of the bathroom and all the other things in this house, which he had made and put here. And that is one of the reasons that I don't want to leave, I don't want to leave all his kind thoughts and work and go away and live somewhere else. I couldn't do it, not unless I really had to. (Mrs Farris, widowed, London)

> When you move ... I don't know whether you've experienced that or not ... but the home isn't yours until you decorate it, and the atmosphere doesn't alter much when you first move into a house, if you don't decorate it before you move in. It takes a lot of [work]. It's everybody else's paper on the wall, everybody else's this. Fortunately, touch wood, we've just about got rid of everybody else's stuff and everything is ours now. We've a big living room here; the kitchen was long in there, but not as wide. So I mean yes, it's good. We can come in and do our own thing in it. (Mrs Otley married, owner-occupier, stay put scheme, Lancashire town)

Mrs Otley talked about the work that had gone into making the home 'theirs'. Similarly, Mrs Farris was attached to the energy, and the 'kind thoughts and work' that have gone into making the house into a place that is physically right for them.

Some people adapt the place they live to suit their needs, an approach that is encouraged by 'stay put' schemes. They may add handrails, move downstairs or alter the garden for lower maintenance.

> *I*: From what you were saying before, essentially your first retirement decision you decided to stay where you were.
> *Mrs Cartmel*: Yes, because the house was built for my requirements, because I had arthritis of the spine. In fact they told me I'd dislocated my spine, but arthritis developed and for ten years I was walking with a stick before I fell and did my hip in. But I stayed in that house because I made the house to suit my purpose. I'd taken the bath out, put showers in. Everything ... extra hand rails and things like this and I was quite content until my wife died. (Mrs Cartmel, 73 years, widowed, residential home, Lancashire)
> *I*: Have you had a lot of modifications done in the home?
> *Mrs Parfitt*: No, not really. The only thing I had was I had the oven taken out, ... and I just had a microwave with convector oven, because I can't bend down to get to an oven. And, as I say, I had the conservatory done. That's all been done specially, and instead of having a lot of garden. ... Of course mine was the biggest garden actually, would have been, but that proved to be better because with the ramp and the turning circle, I wouldn't have got it in the other bungalows, there wouldn't have been the length for it.
> *I*: So you were lucky there?
> *Mrs Parfitt*: I was lucky, yes. And the builders built some raised beds. (Mrs Parfitt, adapted private home, Lancashire)

In both cases, the home has been over time moulded into the requirements of ageing bodies. Doors, showers and gardens have been changed to accommodate the disabilities of age. Bodies, in turn, become more emplaced, able to function within an environment that has been adapted.

The physical characteristics of a house are transformed through the act of living in it. Those who have spent a long time in their home have adapted it to their needs, to make it a better 'tool' for the job of living their lives and reflecting their desires. They have in a sense made it a part of themselves, embodying their values, needs, efforts and lives. In turn, as a home becomes so personalized, the attachments to it become in many respects more profound, reflecting a mutual dependence and therefore often a reluctance to move.

Home as 'doing': daily activities, rituals and identity

Home is also a place of activities and routines. It becomes almost impossible to think about habits, hobbies and interests that have been performed over a number of years separate from the home, the place where they have been undertaken. For example, it became apparent very early on that many older people constantly referred to their need to 'potter'. Of course, interviewees differed in how they defined 'pottering' and in the importance they placed on it, but the extent to which it naturally and spontaneously arose in conversations was striking. Sometimes, they were referring to the routine of everyday activities, other times, to specific hobbies. 'Pottering' seemed to be used as a way of describing how people ordinarily and naturally fill in time, sometimes purposefully and at other times, more aimlessly.

The following extracts show how people found specialist accommodation off – putting because they suspected there would not be the facilities or space to allow them to pursue these activities and interests.

> I think pottering is quite important actually as you get older because I've got a workshop downstairs and I've got a small conservatory and I like pottering in the garden. You know it keeps your interests going. And I think, in a sheltered home you wouldn't have those facilities really. (Mr Black, 75 years, married, 3-bedroomed flat, London)

> *Mrs Keeley*: You know if you're out somewhere or you're away, I mean, I've been on holiday and I thought to myself, why am I here? I'd much rather be at home.
> *I*: Really? On holiday?
> *Mrs Keeley*: Yes, yes, because I think of all the things I like doing you know, pottering in and out the garden and just doing things in the home I enjoy so much really. (Mrs Keeley, 71 years, married, own house, London)

Pottering is in part a process of walking round a house or garden to re-locate yourself in relation to your world. In a sense it is a maintenance of rituals, hobbies and the like that maintain a sense of self-continuity and purpose. The following extract shows how people feel ill at ease when the rituals imposed on them are impersonal and alien. In fact this woman felt so strongly about the issue that she took the trouble to write a follow-up letter, making a passionate plea to be heard:

> One of your interviewers came to interview me a few days ago about your project *Housing Decisions in Old Age*. I was grateful for the

opportunity to take part in this very important research study. Some issues however were not raised and I would like to add one or two points hoping it is not too late to include them as they involve the 'inner life' as opposed to financial and physical problems. Much distress is caused by patronizing and ignorant carers who do not realize the importance of a healthy mind. The tea parties, bingo halls and community singing of First World War songs such as 'It's a long way to Tipperary' or a Beatles number 'When I am 64' are anathema to many of us who were not born until well after 1918 and who are nearer 84 than 64. It is a mockery to sing these songs and many pensioners clubs need to be told that old people may be organized under one authority because of their age or frailty but many of us depend on our lively intelligent conversation, classical music and good literature. To have a paper hat rammed on my head and be unable to walk or talk after two strokes while a young nurse told me it was Christmas and I must enjoy myself – internally I was praying for voluntary euthanasia.

During your assembling of why we are often timid of selling up our homes to live in residential communities please somewhere mention that we are not all ageing idiots and that there might well be less dementia if our tastes in entertainment were considered important. We are not snobs – we just feel 'left out' and not listened to. This is a question of taste, not money. We do not wish to be banged up with totally different IQs than our own. An OAP is not always dreary – the NHS may look after our health but we must keep a healthy mind. (Mrs Fletcher, widow, living in sheltered housing, London)

Home as a social network

All too often, the conception of 'place', 'home' and 'belonging' is attached to the physical, the bricks and mortar of a house, and to a 'territory', or a landscape. However, as well as attachments to the physical essence of places, there are attachments to the people that inhabit these places. People – friends, family, neighbours, familiar faces – may be more important (and so more difficult to leave) than the physical house in which people have spent most of their life. Asked the reason for a move to a particular housing scheme, Miss Barrow replied:

Well, the fact that I'd seen it and thought it was lovely, and the fact that a former colleague and her sister lived here. That was one thing. I would know somebody when I came. And in fact practically

> everybody in the village is an old pupil of mine when it comes down to it, somewhere along the line. That's an exaggeration. But it's surprising how many are. You get a lot of old pupils as you go through life. (Miss Barrow, 89 years, single, bungalow, Lancashire)

In this respect, many feel a definite attachment to a place based on their social embeddedness. They have, over many years, contributed to their towns, villages and neighbourhoods, and see this reflected back at them in the familiar faces, the hellos, and the daily inquiries about family members. Hegel (1977) makes a powerful argument for the role of others, institutions, and laws in the construction of the self. For Hegel, the development of the self is dependent on the action of others, and identity is constructed through the recognition of one's characteristics and traits, as they are seen through the eyes of others.

The three passages below reflect these feelings particularly well. Mrs Fletcher is able to see the results of her long teaching career in the faces of those around her. As a result, when it came time for her to move, she decided to move to a nearby sheltered housing scheme to maintain this sense of belonging and recognition:

> I said I don't want to go too far because I was a teacher here for twenty-one years and I knew many people ... walked down the street, (and got) hellos. Children, I talk to them now, are grown up. Some are even doctors. And I felt at home here. My children have grown up and left, my husband left, so I needed friends and neighbours. (Mrs Fletcher, widowed, sheltered housing, London)
>
> Everybody thought, when he retired we'd move away. But it was not the right time to move, I don't think, not when you retire, you've got your faculties, you've got, well you've got your friends. We've got two sons you know, we are always here for them when they come home. (Mrs Oliver, 71 years, married, 3-bedroomed house, London)
>
> Well, I think. ... I'm born and bred in (Arkholme) as my parents, grandparents and as many generations as you can go back lived in (Arkholme). So it's your roots, and I just thought, 'Well, when I finish work this is where I want to be'. (Miss Davidson, 60 years, single, living in council owned, sheltered housing 2-bedroomed bungalow, Lancashire)

Mrs Oliver was reluctant to move away from friends, and reluctant to stop being the chain that links her sons to the place they grew up. Miss Davidson voices similar feelings, but uses the word 'roots' to express how her identity is bound up within the social networks in which generations of her family have been a part. In all cases, such attachments to place are based on the social, not the physical.

As a last example, we can see how Mrs Bolton, finds a real comfort in knowing that people in her neighbourhood are aware of her and her habits:

> *I*: What is it about here that makes you decide to stay? Why do you say that you'll never go?
> *Mrs B*: Oh everything about it. I like it, I like the people, they are very kind. You see, I always remember once ... I have these curtains drawn at night. In the morning I get up early and I like to get my work done ... . So I don't come in here, I keep the curtains drawn until I've done everything else. Well, it was amazing. I had been doing this one time when somebody came knocking on the door to see if I was all right because the curtains were drawn. I think they wondered if I'd popped my clogs during the night. Well, you see, people notice here, take care and they do something about it. You will never find me being dead six weeks in the house, because there will always be somebody come to find out. And that's a good thing. (Mrs Bolton, Lancashire)

Thus people's attachments to home can be seen as attachments to the social networks that make-up the neighbourhoods, towns and villages in which they live. The testimonies above show how many feel that their place in a locality of social networks is something to be valued. Such valuation is demonstrated in a number of ways: the status it brings, the sense of belonging or roots, the company of friends, or in the feelings of comfort that emerge from living in a community where people feel looked after.

Conclusion

In this chapter we have explored the different and sometimes complex ways that respondents in our sample were attached to their homes. This ranged from the most 'physical' aspects related to the home itself, such as the sense of privacy and control one's 'own' home can offer, or the home as a collection of possessions which give people a sense of narrative continuity in their lives. We also showed how people in a sense feel embodied or emplaced in their homes, meaning that they are both physically and emotionally attached to the ways in which their homes have adapted to, and have become reflections of their lives, tastes and values.

In the last sections of the chapter, we moved away from more 'physical' notions of home to how homes are lived and experienced. Some were attached to their homes as a locus for daily activities such as pottering, and it was those activities that provided people with their sense of who they are. Others suggested that home was important in terms of the series of social networks which had developed as a result of living in a place.

Many people may be torn in terms of their positive attachments towards a place (say, their status and recognition in their village), and other negative aspects of their current situation (e.g., not being able to use the stairs in their house, or the house and garden becoming too difficult or expensive to look after). The result may be to continue to live in an unsuitable environment for fear of losing what they feel is an important aspect of what makes them 'them'.

We could see such a case with Mr Read and his mother discussed earlier in the chapter. They continue to live in a large house that they cannot afford and which is physically inappropriate for them. But the reason they stay, the reason they adapt and modify the home, and the reason they persevere although short of money, is that their home is large enough to contain all the objects and possessions they have collected throughout the years. They do not want to move somewhere smaller and more affordable because they do not want to throw away these things. They feel that it would be throwing away their past family life, and in a sense, part of themselves.

An understanding of the nature of these attachments is an essential ingredient in being able to provide advice that is both helpful and relevant to each individual. Without such understanding, there is a risk of assessing people's decisions without the insight into the reasons why certain decisions were made. Such knowledge of what matters is a key factor in making housing decisions that match what people want.

6
Worried Lives

Feeling vulnerable

For some people advancing age is a time of uncertainty and vulnerability. Often an event such as an illness or a fall is the trigger to the realization that one is vulnerable, a state that may change one's attitude to the whole of living. Hepworth (2000) lists vulnerability and risk as one of the key issues of ageing alongside: 'body and self; self and others; objects, places and spaces; and futures' (p. 9). He writes of the 'risks associated with physical frailty and with particular places and spaces, such as a high crime area of an inner city' (p. 9). In a later section, drawing on Leder (1990) he notes that people live most of their lives 'without being fully aware of the internal workings of the body which, when it is free of pain, disability and illness, we experience as "absent" ' (p. 38).

Ageing results in an increased awareness of one's body, and its frailties. The consequence is that there is a re-evaluation of risks in daily life, in part based on reality and in part on a changed perception of risks. So an incident such as a health scare, which may prove unfounded, nevertheless may make people aware of their bodies and of their vulnerability. The sense of vulnerability may result in people specifically being more careful in what they do, for example, to ensure that they do not fall; but, more generally it may lead to a changed perception of the person and their place in the world.

Thus older people *may be* more vulnerable physically, for example, being more likely to face respiratory problems or to fall, because of mobility problems. In addition, they *may feel* more vulnerable, because of health problems or because of their knowledge of the problems that confront other older people. They may be worried about the present and they may be worried about the future. In addition, they may be acutely aware that their physical strength is considerably less than it was.

Feeling more physically frail, people may feel less in control. Indeed, one of the anxieties of some people is that their independence, their control of their lives, may also be vulnerable. Thus there may be concerns that others, children or professionals, may start to take over decision-making. Hepworth uses Bawden's book *Family Money* to illustrate the point. In old age Fanny Pye suffers a physical attack 'which exposes the vulnerability of Fanny's body'. There are further consequences:

> (The attack) results in a sequence of interpersonal reactions which put her independent selfhood at risk and turn her life into a struggle to resist the efforts of her family to transform her life into one of aged dependency. ... At the centre of this social interaction is a shift in the balance of power operating in the relationship of Fanny with her children. (Hepworth, 2000, p. 103)

Later, Hepworth makes the point that some older people take risks.

Thus, there are factors related to ageing which create problems in daily living and are likely to lead to increased awareness of one's vulnerability. The central factor is that increasingly people have to take account of their bodies in deciding what to do: less physically strong, they are also more likely to have mobility problems or health problems. In general, people are more vulnerable as they age to falls and accidents.

There are two other major strands. The first is anxiety about money and whether, on fixed or declining incomes, they will have enough to live as they want. The second is that people often feel more vulnerable to physical threats, such as crime. Both of these will be developed later in the chapter.

Vulnerability, as we have stated, is based on reality and on feelings or perceptions. Emotions can, and often do, have a cognitive component, as they are, at least partially, based on judgements. For example, people may be anxious that things might happen to them or others which will result in them becoming less able to manage. However, relying on emotions can result in poor judgements, as the event may be unlikely to happen. In this case, the decision made is based on a false premise.

There is arguably a wider cultural issue here which contributes to an increasing sense of insecurity and heightened sense of fear amongst society at large, and to which older people are particularly vulnerable. In *The Culture of Fear*, Furedi (2002) makes the point that, for the last three decades, English-speaking societies have been enveloped in a culture of fear. Fear, he argues, has become less a product of personal experience of specific instances, and more a result of 'theoretical risks' that are increasingly exaggerated by media coverage and political discourse. Thus it is common to report the age of victims in reports of crimes in local papers, for example, that a pensioner was mugged or that police were appalled at theft from a defenceless older person. Yet there is clear evidence that it is young males who are at far greater risk of assault. Through these influences, we, as a society, are made more aware of risk, and therefore more influenced by risks. In this sense, there is a relationship between increased knowledge, risk and fear (Beck, 1992).

For Beck, this increase in knowledge is coupled with a cultural milieu and social structure that has undergone a loss of faith in public institutions and their ability to solve problems or even propose solutions. So this heightened sense of risk is exaggerated by a diminished sense of control over risk, and therefore one's life. One can see how this becomes particularly relevant for older people as this chapter unfolds.

Fear of change becomes a particular articulation of persons in the 'risk society', and fear of change means that people are more likely to look to the future with, if not fear or dread, then certainly caution. This affects decision-making in that fear of the future is heightened by the impossibility of knowing the outcomes of one's actions. Thus the status quo is often preferred to the risk-laden unknown.

Some of the fears described below would seem to confirm this argument, while others can definitely be considered responses to actual personal experiences, as opposed to 'theoretical risk'. However, in either case, what follows is a discussion of the major fears and worries that influenced the housing decisions of the respondents.

Personal experiences of crime

In some cases, older people in the sample have been unfortunate enough to become victims of crime. While such events are traumatic enough for anyone who has the misfortune to experience them, for older people these events tend to occur within the emotional context of 'feeling vulnerable'. A person who already has worries about health or money, and feels more frail and vulnerable, is going to be even more affected by the trauma of crimes. Mr Mercier had an appalling experience. He described how he had lived since 1984 in a Lambeth council basement flat. Tenants had managed to get into his flat and steal £450. He explained that new houses were built with three storeys, and that large families came in, and 'the drug culture sprang up at the other end of the road'.

> And as a result of that, I had about a dozen burglaries, attempted burglaries and two occasions when the boys just burst into my flat and forced me down on the bed and tried to murder me, they weren't trying to kill me, they were trying to keep me quiet while they were looking for money. So of course I always went to the council, there are various ways of moving, and I know more about that than most people. I've been trying to move from a flat where I was since 1983. (Mr Mercier, single, council owned sheltered housing, Lambeth, London)

> I think security is a very important factor: people tend to worry about it. I think it makes all the difference to you, even if you're in sheltered accommodation. I know, because I had a very bad burglary before I went into sheltered accommodation and it's a very traumatic experience, to anybody. I do think it makes you very scared of where you are and I know once I went into sheltered accommodation, the security was the main thing that made me feel better. Which is a pity really. In this day and age, it's awful. But it is a factor, there's no doubt about it. (Member of Newcastle Better Government for Older People focus group)

> In Lancaster, where I lived since 1945, the couple next door to me had died and they sold the house and a deaf and dumb boy bought it. He was great. He was a lovely lad, don't get me wrong, but the trouble was, if I was having trouble with the children at night (throwing stones at window) I couldn't knock for him to come and help me. Then I brought the police up one night and they said, 'Never go out'. (Mrs Denson, 67 years, widowed, 5-bedroomed house, Lancashire)

In the extracts above, the fear of crime is based on actual personal experience, as all three lived in what could be considered problem areas for crime. These fears proved to be crucial in triggering their desire to move to sheltered housing, which they thought would provide more secure accommodation.

Fear of crime

> 'The elderly are allowing themselves to become prisoners in their own homes because of a completely unrealistic fear of street crime', warns a leading charity. Age Concern says a survey of 4,000 older people found almost half of those aged over 75 were too afraid to leave their homes after dark because they believed they would be subject to verbal abuse or mugging. Two-thirds said they believed they would

> inevitably become victims of crime as they got older – while a fifth said this fear had contributed to a sense of loneliness and isolation. The results mirror other research that suggests fear of crime is completely at odds with experience. (Casciani, 2003)

This article by the BBC, referring to a survey by Age Concern, touches on a significant theme in our findings. To no one's surprise, many older people reported a fear of crime as a major reason to move. Such general concern is often driven by the wish merely to have someone around who can be relied upon for protection and help in difficult circumstances. These findings concur with Zedner's (2000) argument that preoccupation with crime has been overlain by a concern for protection and freedom from the anxiety that seems to be an inevitable accompaniment to modern life. As the news article above suggests, older people often suffer from a 'risk–fear paradox' (Jefferson and Holloway, 2000): despite the statistical fact that older people (and older women in particular) are the least likely to be at risk of crime, they are generally the most fearful of it. Jefferson and Holloway think that people's fear of crime is often not a direct response to the risk of victimization, but is mediated by anxiety and setting up defences. Their results undermine any assumption that individuals are solely rational, with their fear of crime a direct reflection of their risk.

The need to alleviate the general anxieties associated with security has weighed heavily in the making of housing decisions:

> But increasingly people didn't go out on their own at night. You know it's crazy. Where we had the shop, there'd been a hold up at the post office a few doors down. It doesn't really make you feel like staying. (Mrs Boyle, married, 3-bedroomed house, Lancashire)

Safety has long been recognized as a basic human need which has to be met before what have been termed 'intermediate level' needs, such as 'belonging and love', 'esteem' or 'cognitive' needs can be met, and well before 'higher-level' needs of 'aesthetics' or 'self-actualization' (Maslow, 1987). The extracts below show how people search for places where their need for safety and security is fulfilled:

> I am now living in a second floor two-bedroom apartment with security entrance. The reasonably priced service charge covers cleaning of communal halls and stairs, gardening and building insurance. I am extremely happy here. I should have made the move years ago. I sleep

> soundly at night now due, to no doubt, to feeling secure. (*Written Housing stories* – Miss Fairclough, 77 years, single, North East)
>
> Well it (*meaning of home*) means a place of sanctuary. It means a place of relaxation and choices. It means being able to close my door and know that I'm going to be safe whilst I'm in there. (Mrs Bosworth, 57 years, divorced, living in 2-bedroomed flat, Lancashire)
>
> That's a local authority one (*referring to sheltered housing scheme*). Now I like that. I think that's good because you know there's somewhere you have a bell connected to the warden. So you know there's somebody there. (Mrs Semple, 88 years, widow, 2-bedroomed house, Cumbria)

These situations differ from the more incident-rated moves of the last section. In these cases, what has propelled a move is a more generalized fear of not feeling safe or secure, an increased sense of vulnerability.

Money worries

A common theme in the gerontological literature, and certainly one of the most notable aspects of post-retirement, is the change in income that most people have to face. The transition from a working life of paid employment to a life dependent on the fixed income of a pension was a persistent theme of respondents. They were especially concerned that expenses may rise as health begins to fail.

> We sold that and bought a flat because I could see that the work, it was too big for us. And we bought a flat, which was a question of maintenance there, general maintenance. And I decided I was coming up to retirement and I said, 'Now look we must consider this, we must think about moving because if anything happens to me, you are going to be faced with bills that you can't meet, not working, for maintenance'. I thought, I know I shall only retire on a limited pension, not a fortune, so I've got to face this. So we sold the flat and bought the maisonette. And the maisonette was freehold. I had the freehold property there, and that was very, very nice until the years rolled on and the wife's illness made it impossible really for us to stay there. (Mr Page, 87 years, married, 1-bedroomed bungalow, London)

While descriptions such as the one above could be seen as straightforward rational planning and decision-making, there is a case to be made for an emotional dimension. Many decisions are best understood by recognizing the intermeshing of concrete and abstract, rational and emotional. In this case there were concerns about the capacity to maintain the former house but there was also an underlying concern, not based on a particular incident, but related to concerns about an unknown future: Mr Page worries about his wife's future ability to cope without him, and initiates a course of action to alleviate that fear. We are not arguing that this is 'non-rational' or 'irrational'; it is rational to recognize that you will die and may do so before your wife. However, Mr Page's worries and actions are linked also to fears of vulnerability and, probably, to their relationship as a couple and his role as provider within that.

Some also stressed their fear of being a financial burden on their children:

> I wanted to be independent. I've got three children, all married with children of their own and not that well off, so I didn't want to be a burden to them. I wanted to be somewhere where I could be independent but be financially secure on a pension and know that my years ahead were looked after and that they needn't have to worry about me. (Mrs Raynes, 63 years, widowed, 1-bedroomed cottage, retirement village, London)

In both types of case, we can refer back to Beck's (1992) concept of the 'risk society': the heightened concern for the future, given the loss of faith in (and power of) public institutions and the welfare state. There is an increased worry that citizens will have to look after themselves because of declining pensions, coupled with doubts that the 'cradle to grave' welfare state will provide for people in their old age.

Worries about mobility and falling

Mobility, the freedom to move of one's own accord, unfettered by boundaries, barriers or hindrances, is taken for granted in our society, until restrictions are placed on ability to move. Mobility is related to power. Those who are powerful have more freedom of movement than those who are powerless, whose movement is often curtailed by social and/or political bodies. At one level, that of movement between states, it is apparent that authorities in Western countries are much less likely to accept the poor or the low skilled into their countries, whereas the wealthy and the highly skilled are welcomed as immigrants in almost all parts of the world (Urry, 2000).

In terms of house movement, in general it is house owners who move more and have greater flexibility in where and when they move. Richer tenants have also got greater flexibility. However, poorer tenants, looking to housing associations and local authorities, become dependent on interpretations of their circumstances and needs by an organization. There are of course limits to the flexibility of movement in any situation: house owners may not be able to sell their property; or the sort of place wanted, whether by tenant or owner, may not be available in the chosen location. The same applies to the ability to stay put. There can be little dispute with the statement of members at one panel that sufficient money would allow anyone to buy solutions to their housing and living needs. They cited the case of someone who could no longer look after herself and lived in an unsuitable house in a village: the family bought in 24-hour attendance. Alternatively they could have paid for specialist housing and specialist care in the place of their choice.

It is important to recognize that money is not the only component of power. Knowledge of what is available and of how to manoeuvre around in different systems is also a key factor, as we discuss in the final chapter.

However, our focus in this section is on an individual's personal mobility rather than house mobility. Restrictions on movement have an immense psychological impact on our sense of autonomy and power. This is particularly true within the context of growing older. Restricted mobility may become part of a person's ageing story, and with it, the worry of dependence on others and less participation.

> You see, when I was mobile, I was very independent and I did everything for myself. I did whatever I had to do. If I wanted to go out anywhere I went out, I didn't have to rely on anybody. But of course, now I'm not mobile I have to rely on people for nearly everything, and it's very frustrating. I mean, I'm very grateful to everybody because everybody helps me; I'm not complaining about that. My carers are very good. I have three carers every day and they are very, very good to me but it's not the same as doing everything for myself. Well, I'm not so much angry, but frustrated because ... I used to have everything in its own place. I knew where everything was. Now I have to rely on other people. And I can look for something and I don't know where it's been put ... and it makes me feel very frustrated. I mean, they are all very good to me but it's not the same. Because when nearly all your life you've done everything on your own without depending on other people, it's very difficult to get used to what I'm used to now. I've accepted it now, I'm philosophical. But

> it is a bit frustrating. (Mrs Sewell, 79 years, widowed, sheltered housing, London)

The frustration is evident. Here a clear link is perceived between mobility and independence. Having once seen herself as independent, Mrs Sewell now feels she has to 'rely on people for nearly everything'. She talks about the real problems of daily living: the difficulties arising from not being able to do your own shopping or not being able to find things.

This was a recognized and common worry amongst the interviewees. Several people indicated that a loss of mobility was an important factor in deciding to move, to avoid using stairs, making long journeys or having to use a car at all.

> We looked in the area of [Lancashire village] because it's flat and it is quite near to shops; we're getting older and, if you can't always drive, you've got to be sort of self-sufficient in that way. (Mr Sopworth, aged 71 years, living in own home, Lancashire)
>
> *I*: [You say] you were looking ahead to when you couldn't drive your car, but you could still stay where you were?
> *Mr Farley*: That's right. We used to see so many houses up in the Lakes and think, 'Oh a lovely to place to live', but you see, there are lots of houses round here. The village is alongside the river. You've got to jump into a car to do anything, and if you can't drive, you're stuck. (Mr Farley, widower, residential home, Lancashire)

Falls have been recognized in both literature on ageing and in governmental policy as major features in the lives of older people. Thus, reduction of falls features as a target in the National Service Framework for Older People (DoH, 2001). Easterbrook and colleagues (2001) note the impact of falls, and fear of falls, on people's lifestyle:

> Even where injuries appear to be minor, perhaps resulting only in slight bruising, individuals can lose their self-confidence as a result. If they then reduce the activities they carry out, their independence can be compromised and more support services needed at an earlier stage (McIntyre, 1999). Fear of falling, and the fear of undertaking tasks or activities that might lead to a fall can lead older people to become disempowered, more isolated, and with a reduced quality of life. In addition, if informal or family carers fear that their older relative or friend may fall, they may be less willing to support or enable the older

> person to take part in activities they perceive as risky (Tinetti and Powell, 1993). ... there is some evidence to suggest that the anxiety generated by fear of falling can itself induce dizziness and increase imbalance. (Yardley, 1998 as quoted in Easterbrook *et al.*, 2001)

Mrs Webster spoke about the problems:

> Well the flat was degenerating, I couldn't get anything done and I couldn't afford to do much myself. I was getting illnesses. I was falling over. I was falling down the stairs and I was on the second floor. And also the building had been gradually going over to commercial use, which meant that at night and at weekends I was entirely alone in a four-storey building; nobody could have got to me. My family was a little bit concerned about it. I mean, I had about forty stairs, no lift and I was falling down the bottom flight. I fell twice and I fall rather easily. I don't have very good balance. I fall in the street. I fall anywhere. (Mrs Webster, 83 years, widow, sheltered housing, London)

The common predisposing factors for developing a fear of falling include being aged over 80, visual impairment, a sedentary lifestyle and no available emotional support (Murphy *et al.*, 2003). Fear of falling leads to reduced physical and social activity as people try to avoid taking risks; however fear of falling itself increases the risk of falling (Lach, 2002/03). Thus concerns about falls become a major factor in precipitating a house move.

'It could be me'

> Well, we had to say the fors and againsts and so we decided there were more fors. My husband's being on his own, which is a major thing, I didn't want to leave him on his own, because to be quite honest ... you hear about ... in the paper, people dying in the houses on their own.... To be quite honest before we left London, there was a lady ... we were young then, we never thought about it then, but there was a lady in the road, she was in the house for a week dead and nobody knew. And her son came up from the country and found her. And when you get older you read all these things and I thought, 'Oh I don't want it happening to us'. And my mother she always used to say 'Oh, I don't want to die on my own', but anyway she died in hospital. I know you get a bit morbid when you get older but you've got to think of these things. (Mrs Carrigan, 3-bedroomed house, London)

Fear of dying alone and being abandoned was a recurring theme. The extract above has an intensely powerful message. Underlying it is the suggestion that many older people are not only making decisions on how they want to live, but also on how they want to die. Several respondents did confess that they had heard of cases either in the media, or through friends, of older people dying on their own and not being found for a number of days. The greatest fear seems to be not of dying on your own but of what is implied by nobody noticing your death. The thought that, 'That could be me' seems to have acted as the catalyst for a number of moves to more communal and/or staffed forms of accommodation.

Independence worries

Many of the respondents spoke of the need to maintain a sense of independence in their lives, and their worries about a loss of independence. The subject of independence forms a powerful background to debates on older people's needs and quality of life. (Secker and colleagues 2003 provide an excellent review.) The overarching theme is that it is necessary to assist in maintaining a sense of independence in housing or home help schemes. The word itself has achieved rhetorical status within government policies and discourses.

Secker and colleagues, following the work of Rose (1999), argue that notions of independence and self-reliance are the result of the internalization of government and psychological discourses on the ideal of being an autonomous, self-actualizing individual (Secker *et al.*, 2003, p. 378). We are, according to Rose, 'obliged to be free', and Secker and colleagues note that this discourse has filtered down into a strong desire for older people to maintain their independence. This is also seen in recent governmental programmes which emphasize self-reliance, choice and independence, for example in 'stay put' schemes. Independence has become a powerful, ideologically laden term, and yet its meaning within the lives of older people in making housing decisions is at the same time prominent and complex.

We have discussed the term 'independence' in relation to dominant ideologies in Chapter 3, noting that Dalley (2002) considers the term fails to portray the lived experience of many older people, and now return to that discussion. 'Independence' is ill defined and used in a taken-for-granted way to refer more generally to a 'lack of dependence' (Secker *et al.*, 2003; Sixsmith, 1986). As Secker and colleagues note, recent literature has begun to deconstruct the notion of independence. Several meanings for the term emerge, such as 'being able to look after oneself', 'not being indebted to anyone', and 'the capacity for self-direction' (Sixsmith 1986 in Secker *et al.*, 2003).

Regardless of its origins, the need to feel that one is autonomous, self-reliant and in charge of one's own life was a dominant theme in the interviews. Worrying about losing one's independence can both provoke and inhibit moves. Most often however, the desire to maintain independence is manifested in a decision to stay put, or at least a fear of moving into some form of specialist housing:

> *Mrs Barnsley*: No, we don't want sheltered unless we have to. I mean we are pretty sane.
> *Mr Barnsley*: I think we would want to be independent.
> *Mrs Barnsley*: Mm, we like our independence, yes.
> *Mr Barnsley*: Yeah, and if you go and live in sheltered accommodation ... obviously there are advantages in it but there are disadvantages that you're not so much of a free agent. And I would imagine, these places with wardens are fairly small.
> *Mrs Barnsley*: No, we value our independence, don't we, until we are forced to think again.
> *I*: So, you would think in terms of surviving on your own as long as possible? That's what you seem to be telling me?
> *Mrs Barnsley*: Yes, yes, oh very definitely yes.
> *Mr Barnsley*: Oh yes, yes, yes, I would go so far as to say that probably this is symptomatic of people of our age group.
> *Mrs Barnsley*: Yes, yes.
> *Mr Barnsley*: That they were brought up to be self-sufficient and independent.
> *Mrs Barnsley*: Self-sufficient and get on with it. (Mr and Mrs Barnsley, married, 3-bedroomed house, Lancashire)

> The trouble is an awful lot of old people, and quite rightly so, are very independent. They might not have any money, but they don't want to be dependent on somebody and they don't want to go in a home. They ... prefer to stay in their own place, which they know, and their own belongings and surroundings. No, if that came to me, not being able to get upstairs easily on my own, I would try and invest or try and get a chair, the stair lift that goes up the stairs, so that I could still go upstairs. I am a very independent person. (Miss Nash, single, owner-occupier, 3-bedroomed house, London)

In both interviews, the assumption is that managing on your own is synonymous with being independent. The decision to move to sheltered accommodation, for example, is seen as a fundamental one in which a person 'gives up' their ability to look after themselves with a degree of self-reliance. Thus the tendency is to stay put as long as possible, as the continued maintenance of private property is seen as evidence of overall mental and physical capability. This attitude is epitomized in Mrs Barnsley's comment 'We are pretty sane' with regard to not moving into sheltered housing. Others frequently assert that they are not ready yet for a move to specialist housing.

However, as we stated earlier, independence is a complex concept, and is used in a variety of ways by people in different circumstances. Very few people considered that they had lost their independence. Instead, people re-define what independence means to them when their circumstances change. In the example below, Mrs Hare is blind, and living in a residential home. While relying on others for most daily tasks, she still believes she is independent despite the fact that others of her age might not see her in that way:

> I think the right answer was to come here. ... I am very independent, as much as possible now. But I cannot thread a needle or do anything at all; writing, reading is absolutely (impossible). ... I go downstairs, I go down to a friend's, this was one dear friend when I lived here, she read all my letters and helped me and signed papers. ... I mean you've just got to have help around. (Mrs Hare, Weston Super-Mare, panel member)

With finer tuned understandings of independence, it can be seen that a move was seen as a way to enhance independence. In the extract below, Mrs Webster shows how she began to reconsider popular concepts of what it means to be independent:

> I don't think I'd have any alternative. I think that you've got to be sensible. I mean some people like to cling on to their, they call it their independence, but nobody thought about independence as much as I did. That was one reason I never thought of coming into these places earlier. But there comes a time when your independence doesn't do you any good. (Mrs Webster, 83 years, widow, sheltered housing, London)

Going even further, Mrs Sewell provides a more considered view of what it means to be independent. Here, she shows that a feeling of independence is not necessarily related to staying put and being reliant on neighbours; she chose to move to maintain her independence:

> Well I couldn't keep on the house; I couldn't keep it on with three pianos. And it had been very busy and it would have been very dreary on my own after all that activity in the house. So I decided that I wanted to be independent. Now people say if you go into a home you lose your independence but I don't agree because you're being looked after by professionals and you are independent. Whereas if I had stayed in the house I would have been dependent on my neighbours and friends and cleaners and gardeners and so on and I wouldn't have liked that at all. And imagine the evenings on your own and waking up in night and feeling rather ill, what do you do? (Mrs Sewell, 79 years, widowed, sheltered housing, London)

For her, independence meant having available a set of arrangements, delivered to a professional standard, to help with daily living. She wanted to be guaranteed a certain quality of service, available in sheltered housing, but not at her former home.

Mrs Walker states that the decision to move to a residential home was 'absolutely my own':

> I didn't even tell the family. I was determined that any decisions were going to be mine and. ... I wasn't going to be put out, put it that way. I didn't want to feel that someone else had got rid of me. ... I think it was playing safe, because I could have dropped and broken my leg. I've had a couple of falls. My plans would have gone then, they would have been taken away from me ... if I'd have been put in hospital ... (Mrs Walker, 86 years, single, living in residential home, Lancashire)

Mrs Walker's move was prompted by a fear of a potential loss of independence. She wanted to make a decision on her own, an independent decision, without either having family make it for her, or an injury forcing her hand. In this situation, her desire to make a decision before circumstances, or other people, decided for her, was a way of asserting independence and keeping to 'her plans'.

Losing loved ones

The previous section highlighted the desire to remain in control of one's own life. However, housing decisions are also taken within contexts of the interdependence that is developed between people in long-term relationships:

> But everything was hunky dory, we were doing all right until my wife died and then that's when suddenly I was in another era, on my own. And I was thinking, 'Well what can I do? I'll try and look after myself'. But I couldn't. There was no argument about it, I couldn't look after the house and go shopping and doing all of that because I'd not done that sort of thing before. And I fell down a flight of stone steps and did my hip, did my leg. (Mrs Cartmel, 73 years, widowed, residential home, Lancashire)

This illustrates a very common theme. The loss of a lifelong partner is one of the major events which contributed to a decision to move. Sometimes the loss of a partner highlights the symbiotic nature and mutual dependence, which is sustained in a long-term relationship. Mrs Cartmel realized after the death of his wife that he was very dependent on her in terms of domestic work and maintenance and was unable to take over these tasks to look after himself. That, in combination with an injury, led to a move into a residential home where domestic needs could be met.

Mrs Rake had moved to Spain with her husband. He died. She realized that she had relied on her husband to communicate with Spanish officials. For her, this made even mundane tasks difficult and, combined with safety issues, pushed her to move back to England.

> *I*: So you then had to make a decision to return to England?
> *Mrs Rake*: Yes my husband had died and I had to make the big decision whether to stay there or come back. And things were a bit difficult, I could speak Spanish but when it came to going to official offices I couldn't manage it, it was that sort of thing. Then I was mugged on the doorstep of my flat there, so er ... .
> *I*: Really, this is when you were on your own?
> *Mrs Rake:* Yes this is when I was on my own. And other little things, like my suitcase was lost at the airport and I was locked out of the flat by the wind blowing the door shut and had to get a locksmith and so on. All these things piled up of course. And so I eventually decided to come back. (Mrs Rake, 71 years widow, 2-bedroomed flat, Lancashire)

While in some ways this is an extreme example, it is not at all unusual for this sort of mutual dependence to exist. Many women of this generation had little experience of earning an independent income, or managing financial activities such as writing cheques, paying bills and arranging for repairs to the home. When left on their own after losing a partner, these tasks seemed somewhat daunting and left them vulnerable to exploitation. In such circumstances, moving may become the easiest way to avoid such issues:

> After I married I didn't work for money very much because he preferred that I didn't, but I did odd jobs occasionally when people were sick and I did a lot of voluntary work with the Adoption Society and youth work as a magistrate. ... Then he died in 1982. Both boys were away from home and married and I was living in a largish family house and decided that probably I should begin to consider something different. (Mrs Cassidy, 90 years, sheltered housing, Lancashire)

These sorts of examples highlight the direct impact of losing a partner on a decision to move.

Worries of status and personhood

In Chapter 1 we have noted Mrs Bennett's comment that before moving to sheltered housing, she was treated like a normal person, doing everything for herself. Hegel writes that individuals are 'recognized and treated as a rational being, as free, as a person' (Hegel, quoted in Douzinas, 2002). Of fundamental importance is the recognition by others of ourselves as free and autonomous, as individuals and agents with a will and purpose. This is of course a mutual, inter-subjective process. To be recognized one must also recognize others.

By contrast, lack of recognition (or mis-recognition) undermines a sense of identity in that it projects an inferior or defective image of self (Douzinas, 2002). One ceases to feel, for example, like an individual, or a being of worth if not recognized as such by others. The authorities in Mrs Wexford's sheltered housing scheme appeared to refuse to treat her as a rational or capable individual.

These sorts of scenarios underlie many people's fear of specialist housing and residential homes in later life. They worry that they will not be treated as human beings. But fear of the loss of recognition is more complex than this. Our society closely associates identity and status with both ownership of property and occupation. For many retiring from paid employment, is difficult enough, but, when combined with giving up property and moving into communal or specialist settings, it may be seen as a real threat to autonomy and recognition as worthwhile human beings. People want to remain 'somebody'.

This desire to continue to be somebody merges with a desire to maintain a certain status. Many of the interviewees valued the fact that they had a certain status in their local communities from having lived in the area for an extended length of time. To move from localities where their status in the community had been cultivated over a number of years would entail loss of status.

> When he first came home, we were offered (a new location) but we said, 'No' because we own this house you see, so we thought we might as well stop here. ... We have been here all this time, so we might as well stop here now for the rest of our lives. We've got all the neighbours round us, so it was just one of those things. If we moved, and owt happened, you are on your own, there's no ... You've to start all over again. It's not very good. So we've got neighbours next door that know us and half the village knows us anyway, because my husband was born here. And I've been here 25 years so I'm counted as a regular now and not an off-comer. For 25 years I was an off-comer. (Mrs Osbourne, 77 years, widow, living in council owned bungalow, Cumbria)

Mrs Bourke's story illustrates different aspects of people's concerns to establish a way of life that suits them. She had been living in her current house for 15 years, and describes the process of moving:

> *Mrs Bourke*: I'd been living for about 20 years in Cockermouth and I'd been a district nurse and health visitor and when I was 50 I got married and he has got family in East Lancashire and I still had a sister in Southport at that time, and we felt out on a limb in Cockermouth.... And this house (where we are now) belonged to a friend of mine and... she knew we'd been wondering about moving further south and she said, if we were interested the house would be up for sale. And I, actually, I did not want to move; I'd been so long there. But then one day my husband said, we were just going down to Lancaster, we'll just do a detour to Kirkby Lonsdale and go past and see what it looks like and of course, it is in a lovely position and so on and gradually, I came round to the idea. I must say, I was very homesick for a long time and missed my friends. I just knew everybody in Cockermouth.... With just being one of the healthcare team

> there, you know most people. Most people knew me. I suffered from quite a loss of identity when I came here. It was a funny feeling. Good for me, I would think. ...
>
> *I*: What contacts have you made in the area?
> Mrs Bourke: Well, church has helped. The first Sunday we went (there were) an awful lot of visitors who were on holiday, so for quite a few weeks, they just thought we were visitors. I did make myself known to the minister, after a few weeks, so that was a way into getting to know a lot of people....
> *I*: Are you involved in any societies?
> *Mrs Bourke*: The gardening association and I think my husband is about to join the Civic Society, but several Church connected things.... And I get involved in a lot of things that way....
> Mrs Bourke: No, I wouldn't want to uproot again, no, nor would my husband actually. It's been awfully good for him coming here because in Cockermouth, he, poor man, was known as Nurse Bourke's husband, which was dreadful for him, whereas here, I'm Mr Bourke's wife and people know him much better than they know me.... So that's really good for him.... It's a very caring sort of community. My husband was one of the co-ordinators of neighbourhood watch, and people are still inclined to come here. (Mrs Bourke, married, own house, Cumbria)

Mr and Mrs Bourke were considering moving nearer the town centre because the garden might become difficult to manage and they were uncertain about the impact of their neighbours' planned extension.

As in the discussion of social networks and roots in the last chapter as representing a potent attachment to place, in the examples above we can see that worries about losing status, social ties and even identity play a large part in a decision about moving. People worry that a move could jeopardize their social status and their autonomy.

Worries of family members

The fears of relatives also seem to play heavily in many housing decisions, whether directly, though strong family opinions about moving, or by proxy in that the older person wants to alleviate the anxieties of their relatives.

> Then I became very ill and I nearly died and my sons were frightened. They said, 'You must live here where there is a warden, and pull

> a cord if you want an ambulance, with the doctor just round the corner'. And they found the flat and they moved me. I was in hospital. When I came back everything was ready. The bed made, the fire on and a lady to look after me for 6 weeks (*paid best friend to be live-in carer*). But I couldn't stand. I couldn't walk. I came in a wheelchair. (Mrs Foxton, widow, sheltered housing, London)

In either case, statements such as the one above point to moving as a response to others' fears, especially when relatives live some distance away. Thus the worries of others are sometimes a persuasive factor in a decision to move.

Conclusion

Our purpose in this chapter is not merely to dwell on the negative aspects of growing older, but to demonstrate the way housing decisions are made within contexts that are highly charged with emotional issues, issues that include worries, losses and concerns for the future.

We have also tried to demonstrate that these emotional contexts blur the boundaries between rational and emotional decision-making. Planning, for example, for future housing needs can be looked at simplistically as a rational process, but what if the planning is motivated by an exaggerated fear of crime, a general feeling of vulnerability, or a fear of dying alone? At the start of this chapter we referred to 'theoretical risks', those that are possible but not probable.

As we have seen, often there are very powerful emotional factors at work during these times of decision: the losses of lifelong partners, declining health, stairs that start to become potential sources of life-threatening falls, the fears of losing independence, autonomy and even personhood.

While 'worry' and 'fear' in themselves may be considered vague terms, we have tried to demonstrate that such concerns may be instrumental in triggering later life housing decisions. This is not to suggest that older people's lives are dominated by worry and fear, as clearly this is not the case for the majority of older people. What we suggest is that fears, worries or anxieties of what could happen in the future, sometimes based on specific instances, are one of the key factors in both deciding whether or not to move, and also deciding where to move.

7

The 'looking glass self'

Introduction

In this chapter, we argue that making housing decisions with which people feel comfortable will partly depend upon people knowing themselves and using that knowledge to decide whether, when and where to move. We give examples of people who judged whether housing options suited their personality, would allow them to pursue their chosen lifestyles and would enable them to keep in close contact with people who make their lives meaningful. In terms of lifestyles, people spoke about significant routines, hobbies and activities.

Knowing yourself also depends on being aware of your own and your partner's strengths and limitations. We discuss the various strategies older people use, both problem-focused and emotion-focused, to cope with the dilemmas which face them as they grow older and the way that affects their housing needs.

Using these insights into one's own needs and goals in life, and into the way housing and location fit into those requirements and aspirations, helps people assert themselves more clearly and firmly when discussing their housing situation with others. The extent to which people feel in control over their housing decisions is discussed in the last section of this chapter, as this has been proven to influence people's sense of well-being and level of satisfaction with their decisions.

Reflecting upon what is important

> I think it was a learning period, a learning period to get to know myself really. So it is a long process of learning about myself I think. (Mrs Panton, 83-year-old widow, residential home, Cumbria)

George Herbert Mead (1934a,b), who developed a theory of a social self, argued that human consciousness is an awareness of self in relation to others. He saw the social self as producing, not produced by, human consciousness, which he considered to be a *social* consciousness. He developed the image of the person as a *reflexive self* – a self that is able to observe, plan and respond to one's own behaviour.

His portrayal of self is one where we have the ability to know ourselves, to be self-directed, able to judge, think and gain insight into our own behaviour, and to be shrewd and intuitive. This image is conveyed effectively by Cooley's (1922) term 'the looking-glass self' (cited in Rogers, 2003, p. 233).

More recently, the notion of reflexivity has been discussed by a number of theorists including Giddens (1990), Lash (1994) and Beck (1992). All of these argue that contemporary society is being characterized by increased reflexivity of all kinds. This ranges from society at large to the individual.

Giddens (1991) uses the term reflexivity to refer to the increased use of knowledge of present situations to organize life. When applied to individuals, reflexivity refers to the way in which people, instead of merely being given a taken-for-granted identity through social structure, can construct their own identity, in a sense, seeing themselves as a reflexive project. Thus, instead of being constituted through external factors, as the previous discussion of 'recognition' would suggest, individuals in contemporary society attempt to construct themselves through constant reflection about themselves and the sets of options they see around them: they work and rework not only their own biographies but also their projections of their life narratives into the future.

This 'reworking of biographies' can be seen, for example, in the ways by which older people changed their views and definitions of 'independence'. The overall point is that through lifestyle choices, individuals sustain, a story (or perhaps multiple stories) about themselves. They do things, pursue activities and make decisions based on their reflexive definition of self, that is, the kind of person they want to be and would like to be seen as by others.

In a similar vein, Gardner and colleagues (1996) have argued for the existence of *intrapersonal intelligence,* the ability to understand oneself and use that information to manage one's life. Of course, as with other forms of intelligence, people clearly differ in their ability to understand themselves and the extent to which they use their self-knowledge when making important decisions. Sometimes, professionals can help people do this more effectively. For example, career development advisors use personality tests, pen-profiles and interviews to help clients think about the careers to which they might be best suited.

Personality

We noticed that when people reported feeling relaxed and pleased with their housing decisions, they often justified their choices by describing how they took into account the aspects of their personality that they needed to satisfy to be happy. In the interview extract below, Mrs Raynes talks about being outgoing, sociable, enjoying the company of others and having many friends, needing people to talk to, being busy and actively involved in the local community – all characteristics of an extrovert, though she does not use this term. She also takes a pride in being self-sufficient, in control and independent.

> *I*: How long had you been planning this (*move to a retirement village*) or had you just been thinking, as you got older you didn't want to be ...?
> *Mrs Raynes*: I wanted to be independent, yes. I've always enjoyed being in charge of my own life. I'm the eldest of eight, myself, my brothers and sisters. And there are one or two who don't want to be in charge of their own life, they live on benefits or things like that, they found it difficult to hold jobs down. But I am somebody who enjoys going out to work, being in charge of my money, life and so I would have found it very difficult to be reliant on my children in my old age.

Later

> *I*: With hindsight, I know you've not been here that long, but do you think the decision that was made was the right decision?
> *Mrs Raynes*: For me, yes. Not everybody takes to the village, I've heard that. And lots of people move out. You can move out of course, you can move out. For me it was what I wanted. I wanted a cottage and I wanted to be in the countryside. But I would have been extremely lonely if I'd had one of these lovely little cottages with thatched roofs and roses round it, down a lane, because I'd have no neighbours and friends and I would have been very lonely. For me, I am somebody who has to be part of the community as well, and so I have got friends all around me and because we're all in the same boat here, we all look after each other. Its not as though I'm a stranger coming in to, well, it's a very established community. Everybody has arrived

here after they have retired. So everybody is aware of the need to look after each other. (Mrs Raynes, 63 years, widowed, 1-bedroomed cottage, retirement village, London)

Many of those who lived in communal living schemes thought it extremely important to be seen to be a good mixer. Some people who appeared to have been able to manage communal living successfully acknowledged that they were probably suited to it in the first place because they had naturally outgoing personalities or recognized the importance of 'mucking in'. Others appeared to manage it because they were prepared to learn how to adapt to their new circumstances.

> Oh yes. It's not the perfect environment ... you're not living in a detached environment, you are living as part of a community and sharing a building. So, you have to recognize that. It's no good going in thinking, 'Well, I'll live quietly in my little corner', because noise can intrude if you allow it to. Some people do get fanatical about noise. (Mr Smart, 62 years, sheltered flat, Lancashire)
>
> Oh yes, yes, I'm not against that at all. But it's 'mucking in' or joining in, I say mucking in because it's a wartime expression, that's the best thing you can do. You've got to do it. I think it's only fair to join in. You shouldn't come to a communal place unless you are prepared to join in. (Mrs Brown, 82 years, married, living in sheltered housing for a month, London)
>
> No, I don't think so. I'm very happy here. It's up to you what you do, whether you just shut your door and stay on your own or you go down and you mix. I think if you live somewhere like this you've got to be able to take part in the community, otherwise it doesn't work. You see, people all go downstairs in the morning about eleven, someone makes tea and we all sit and chat. I don't always do that, I've got other things to do, especially when there's two of you, more so. And then, in the afternoon, they do the same, so people all get to know each other and they are all very friendly, there are only about two who don't come out. Everybody else mixes. (Mrs Dickenson, 72 years, married, sheltered housing, London)

Many people living in very sheltered housing said they enjoyed having the private living space of a flat combined with shared communal living space, as this offered the best of both worlds. Sometimes though, they still described the way that this arrangement effectively came at a price.

Not everyone finds it easy to adjust to community living, whether that is a retirement village, sheltered housing or residential home. When people are more introverted and regard themselves as quiet, retiring, reserved and prefer their own company, they often report disliking the intrusiveness of living closely with others and found ways of opting out. They experienced the expectations and pressures to socialize as a strain and in some cases, reacted to this by leaving entirely:

> One point we should make, that I feel very strongly is that you've got to make a very positive decision that you are going to move into a place like this. You've got to analyse the situation properly and decide. You've got to do the two lists – the benefits and the costs – and again, the ones that have moved in quickly, one couple came to Morecambe for the holidays and thought it would be a nice idea to live here. Within weeks of moving in, it was on the market. Whilst the husband was fine, the wife couldn't cope with the proximity of other people. (Mr Saunders, married, sheltered housing, Lancaster)

Others cope by making sure they have a life outside:

> And my brother lives in Surrey as well. I've got a life outside the village. And of course I've got my children and my grandchildren that I go out to stay with and they come here. There is another life other than life in the village. Now some people apparently make their whole life in the village. I think at the moment I have a bit of both, like I went to the theatre last night with the village, I belong to a Theatre Club. There are all sorts of clubs and activities here that you can join. (Mrs Raynes, widowed, 1-bedroomed cottage, retirement village, London)
>
> We felt that an important factor is being able to keep in touch with our previous friends. We have been members of a small church for thirty-five years and we have been fortunate in having friends here in (place name) who attend the same church and are happy to take us to church services with them. This means quite a lot to us. (*Written housing stories* – married couple, living in sheltered accommodation)

The overall point is that those who have managed to reflect upon themselves, their personalities and what they feel is important to retain in their lives, have in general ended up feeling positive about where they live.

Lifestyle aspiration

> 'Lifestyle, choices and identity are the key concepts of post modernity. (Heywood *et al.*, 2002, p. 28)

During the 1980s and 1990s, much attention was given to the notion that consumption and consumerism had become the pre-eminent form of identity construction in advanced industrialized countries (Baudrillard, 1988; Bocock, 1993; Lash and Urry, 1987). This entailed a shift from identity forms that were based on one's role in the productive process (for example, class) to one's role in consumption and use of leisure time (such as, goods and lifestyle choices) (Bauman, 1987). In other words, one's value as a person became increasingly judged not on what you did for a living, but on taste, style and material success.

Chaney (1997) makes the point that lifestyles have become more significant in modern society due to an increasing lack of rigid social distinctions, more social mobility, and an increasing obsession with 'things' or material culture which are used as means of creating our selves. Bourdieu (1984), by contrast, saw lifestyle as a set of social and cultural practices, which serve as a way of establishing differences between social groups with certain material circumstances.

However, Chaney (1997) emphasizes the dynamic processes through which symbolic meanings are built into the structures and forms of everyday life, arguing that lifestyle is more of a discourse which negotiates the self as a life project (as in our previous discussion of 'the reflexive self'), focusing on appearance, self-definition and the energy to 'create' or present yourself to others.

In any case, there is a strong link between lifestyles, consumption and identity. Lifestyles are concerned particularly with visual appearance and the energy put into constructing the self as a project. For example, pursuits of leisure or keeping up an appearance are forms of identity building. The effort becomes that not only of consuming, but also of consuming in a way that fits in with the strategies needed to build a form of identity that is seen as desirable.

However, these cultural movements had mostly been explored in relation to young professionals with little reference to older people. Gilleard (1996) is the exception. He emphasized that there is a new generation of older people who, because of private and occupational pension schemes, have the means to engage in a consumer-led retirement lifestyle, and thus forge their new post-retirement identities on the basis of identities as consumers. Thus, instead of viewing retirement as the end of productive working life, one can see retirement as the beginning of a consumer-focused lifestyle. This can be seen, for example, in promotional literature for American retirement communities. These emphasize to potential buyers the notion of an active retirement and promote the concept of the 'ageless self' (McHugh, 2000).

It is really this younger generation of older people, employed through the 1980s, who are used to being members of a consumer culture (Gilleard, 1996). Better off older people are now confronted with a range of options to promote their notion of a healthy, active and independent old age (Ylanne-Mcewen, 1999).

In our interviews some older people moved house to pursue leisure activities and interests, such as fell walking or cultural pursuits. Others were attracted to the idea of having an active lifestyle and either moved or stayed put to achieve this ideal.

> What I think, it's because I haven't this sort of attitude, I don't want to spend the last years of my life mortgaged to bricks and mortar when I want to do other things. And I found when I was living on my own, doing the chores and looking after the garden and shopping and cooking, I just didn't have time or energy to do anything else. (Mrs Wells, 81 years, widowed, residential home, London)

> *I*: So at the moment, do you think you are planning to stay here for the foreseeable future, for the rest of your life in fact?
> *Mrs Henley*: North London is very, very expensive to live in, far more expensive than our pensions allow, but it's very difficult. Logically, we'd move in a westerly direction, assuming our daughter would move from south London to an appropriate reachable place. But if you've been used to living in a town all your life, and we're both Londoners, you can't just plonk yourself in the middle of the country because it's cheaper. So you need to find another town which is both affordable and has the kind of cultural pursuits that you're used to or have enough money to keep on coming up to London for weekend breaks like my godparents did. (Mrs Henley, 64 years, married, living in own house, London)

> We've been very happy since we moved. It's lovely not having to work every day and we can have a go at doing all the things we didn't have time for previously, like painting, sewing, walking and attending a few computer courses. I have also bought a computer, which I love and am completely fascinated by it. My husband also finds

> plenty to do; he goes to a club twice a week, does a lot of walking with the dog and model making. We take the grandchildren out kite flying and also find time for grandchildren minding when needed. We tend to share the housework; there are things we each prefer not to do, and so the other one does it. Mostly it works out very well. We are now both in our middle sixties. (*Written housing stories* – Mrs Driver, 65 years, Sunderland)

Housing choices are not only driven by a wish to pursue a certain lifestyle; they are also dependent upon how people are able to deal with the dilemmas that face them as they grow older. We now move onto describing the strategies, tactics, skills, resources and solutions the respondents said they used to cope with the pressures and tensions they faced. Individually, some of these pressures might not have been sufficient to trigger an immediate move, but they did create a feeling of dissatisfaction and sometimes despair that might lead to a later move if other reasons surfaced.

Coping strategies

The stress-threshold model, introduced in Chapter 4, explained older people's housing moves in terms of stress, which for the purposes of the model, is defined as a discrepancy between actual and desired living conditions. It predicted that older people would only move when their perceived stress exceeded a certain threshold level. In this model, stresses were categorized as either *internal*, for example, where a person's needs or preferences changed, or *external*, such as decrease in the provision of local transport.

This model however, underplays the coping strategies older people use to deal with the stresses they face. Lazarus and Folkman's (1984) widely accepted definition of stress is helpful:

> Psychological stress is a particular relationship between the person and their environment that is appraised by the individual as taxing or exceeding his or her resources and endangering his or her well-being. (p. 19)

Individuals differ in the reasons and the extent to which they perceive a particular situation to be stressful and, further, differ in their timing of when a move might be the best way of reducing stress. In making their assessment, they will need to decide whether the situation is potentially harmful and, if it is, whether they have the coping strategies to deal with it. They may also have to reappraise the decision to see whether their coping strategy was temporary, long term or permanent.

Coping strategies can be problem-focused, such as attempting to manage or alter a problem situation, or emotion-focused, for instance, trying to view a situation as less threatening. Folkman and colleagues (1986) have identified eight distinctive coping strategies:

(i) confrontive coping
(ii) developing plans to solve problems
(iii) distancing
(iv) self-controlling
(v) accepting responsibility
(vi) positive reappraisal
(vii) escape-avoidance
(viii) seeking social support.

Problem-focused coping strategies

Confrontive coping

When things went wrong or were not satisfactory, some people decided to tackle the problem head on, fought for what they wanted and tried to get the person and/or organization responsible to change their mind. Sometimes this was successful, other times not.

> When I came here, I came in to be part of the family. After all, fellowship with one another is what I think we should be aiming for. Our managers give us a meeting when we ask. They listen and give an answer; if either 'Yes' or 'No', they give the reasons. (Mr Garrett, 87 years, very sheltered housing)
>
> We used to have a resident warden but she left three years ago; the company said they have advertised, but no response so we have a warden that lives about 2–3 miles away and comes in Monday to Friday, 9 a.m.–1 p.m. When we complained the company told us that the only thing we could do is look for somewhere else to live. We pay £60 per week for no amenities whatsoever. (*Written housing stories* – Mrs Wexford, Bristol)

Importantly, some people were feeling frustrated and confused about where and how they should make complaints about sheltered housing wardens and managers. They did not know how to question the working practices of these managers, who to go to or how to ensure that quality checks were made upon the ways their schemes were operating.

> Can you tell me if there is anyone we could go to who would help us to check up on his or her bookkeeping? We do have a meeting with the housing association, but when we query things, the usual reply is, 'I will check up on that and let you know', but we never hear another thing. So it is not the accommodation, it is the conditions, which I think, need looking into. (*Written Housing stories* – Mrs Erwin, Richmond)

> In spite of my numerous complaints (I have not listed them all!) I still think this type of accommodation is ideal for retired people but there has to be some control over the people who manage these places. (Biographical stories – Miss Erwin, 72 years, single, Cardiff)

Developing plans to solve problems

Some people like to plan ahead, be organized and to anticipate potential problems; they are proactive. Table 1.1, in Chapter 1 showed that 55 per cent of questionnaire respondents either 'agreed' or 'strongly agreed' with the statement, 'I wish to plan now for possible problems which may affect my ability to stay put'. Indeed 72 per cent of respondents 'agreed' or 'strongly agreed' with the idea of being proactive: 'I would consider moving before things became difficult, to avoid having to move in a crisis'.

The solutions are varied and range from considering adapting the house, getting in extra support, making contingency plans for coping with future illness or disability and moving before crisis occurred. Older people were often alerted to the potential problems that they might face, as they grew older through witnessing their parents' housing and health difficulties. This gave them practical insights and a ready-made checklist of what they should look for when choosing their own accommodation.

> Perhaps I was lucky in having to care for and nurse my old parents until they died at the ages of 85 and 92. This gave me valuable experience of our needs as we get older and it made me realize what I needed for myself. I had to move house because I was living in an isolated country position, not suitable for one person getting older. I was looking for my own front door and 24-hour help available (in person not a life-line type of help) when I need it. Also, if possible, a site within walking distance of shops, doctor, dentist etc. I needed no help in deciding what I wanted, but of course, I did have the help of my experience. I most certainly made the right decision in moving

> here. Whilst I miss the solitude of country life, I recognize that I have exchanged it for something precious at my time of life. This type of retirement home should be available to everybody who wants it. (*Written Housing stories*, Mr Connelly, living in privately owned sheltered housing flat)

Other times, they drew upon knowledge gained directly from their own experience to compile a checklist of what they would look for in terms of space (numbers and size of rooms, outside facilities or garden), inside facilities, privacy and security, views and location, and access to support services. These are discussed further in the next chapter.

Emotion-focused

Positive reappraisal

Faced with major disruption in their lives, one strategy people used was to re-evaluate their priorities, values and goals, in short, to rediscover what is important in their life. This was often a turning point in their lives, the moment they saw the world in a different way. Realizing that the desire to free up time and energy for themselves was becoming more important than staying put prompted some people to move and helped them accept the decision was a wise one. They described how, with increasing age, they had less energy so that simple household tasks took longer to complete. Consequently, they thought it important to conserve their physical resources so that they could devote time to more pleasurable social and leisure activities.

> *I*: When you were on your own, after you were widowed, you had a flat?
> *Mrs Waterford*: Yes, yes, yes. And it got just too much you know. And I hadn't got the sort of income that I could afford someone to do the garden or look after me. So, moving seemed to me a sensible thing to do because I have everything done for me here, you know, absolutely everything.
> *I*: What do you mean by everything?
> *Mrs Waterford*: All your meals, your laundry, your cleaning, well absolutely everything is done for you. There aren't any sort of mundane chores that you have to do. So the point is, it leaves you free with energy to do other things, ... to visit friends, to go to concerts and ... I go up to the Royal Academy. I live in a sort of comfort as it

were. That's what I like. And you have companionship if you want it or you can shut yourself away if you want it. That is what I find is the advantage. (Mrs Waterford, 81 years, widowed, residential home, London)

Many people set out determined to be seen as independent and self-sufficient, but found, as daily tasks and activities become more difficult, this desire had to be sacrificed if staying in their current home was to be the priority. Reluctantly, they asked family and neighbours for help.

I: How will you cope with that large front garden? I'm thinking about later on if you can't manage it.
Mrs Armley: I found that the lawn is one long lawn. My mower is a rotary, electric but it's blades, so it takes me quite a long time and I mean a long time. It takes me all afternoon. If I start in the morning about 11 o'clock, I can be going till now and come in round about 5 p.m. But I will have to get help and I'm going to have to ask for a bit of help. I need voluntary drivers too, mainly, because I've got to go backwards and forwards to the health centre. I've got to contact voluntary services. I'll have to pocket my pride. (Mrs Armley, 77 years, widow, council owned bungalow, Cumbria)

Alternatively, others in a similar position may choose to retain their sense of independence by re-evaluating what it means to be independent, choosing to be reliant upon professionals rather than neighbours and friends, as discussed in Chapter 6. Another option is to accept that independence may not be as important as being safe or having people close by to look out for you.

In contrast, others decide that, in the face of mounting difficulties with managing the home or garden, they would give less attention to keeping their garden neat and house clean.

The garden in fact has represented, it's a good representation of our outlook. We do not have neat tidy borders now; we've gone more over to perennials so that it reduces the amount of work that's got to be done. The house is not as clean and tidy as we would like it to be but we accept that we can't do it all now. And we have discussed what we would do and we have decided we'll stay put because the stairs aren't steep, the house is comfortable. I wouldn't like the house just to go to rack and ruin or the garden (but they) needn't perhaps be at the good degree they are at the moment. I could possibly buy in somebody that would come in on a monthly basis just to keep it down. But if it really got away from me then I would have to consider moving, preferably into a self-contained flat. I wouldn't want to go into care unless I was absolutely obliged to. (Mrs Hartley, 79 years, married, own house, London)

Accepting responsibility

Some people saw taking the initiative and making an early move into more suitable accommodation as a sign of independence and a reason to feel proud. Time and time again, people stated they did not wish to be a burden on their children and these next extracts show the importance attached to taking preventative steps to avoid the chances of this happening. They were very aware that their children had their own lives to lead and that having a dependent parent would place them under a great deal of stress. They did not want their children having to travel great distances to look after them, having to spend time looking after them when they were unwell, or having to do errands and chores for them.

I: Was anybody else involved in your decision?
Mrs Pinter: No, only myself. Because having been a widow since I was 40. I made my own decisions, and I discussed it with the family. My son thought it was fine if I wanted it, as sons do. My daughters thought, my elder daughter who used to be very proprietorial, thought, 'Mother, you can't do that'. But I think I've gone up in their estimation since I got older because her contemporaries have elderly mothers who lean very heavily on them and they feel very responsible for them and they're a bit of a blight some of them. And she does not have that with me because I'm here and I am self-sufficient and we just have a very good relationship. She doesn't have to worry about me. No, nobody has to worry about me. (Mrs Pinter, 83 years, widow, own house, Lancashire)

I: Were you, did you by any chance think that if you consulted them they might say, 'Stay here mum we'll get some adaptations made'?

Mrs Warren: No, because I've always been my own person and I live and die by my own mistakes. I thought if they say, 'No, stay mum' and I'm unhappy I'd be blaming them; if they say, 'Yes, move mum' and, likewise, I'm not content, then I'd be blaming them. (Mrs Warren, 88 years, widowed, own house, London)

Self-controlling

Some people preferred to keep their feelings to themselves and not tell others how bad things were. Some interviewees described the ways in which they would keep private their worries about their own financial situation because they did not want neighbours to know of their struggles and economies. There were occasions in panel meetings or interviews when people spoke of distressing events that they had mentioned to nobody else: family feuds and battles with neighbours or a wife having to cope with trying to wash and get to bed her husband who had soiled himself downstairs when he collapsed and could not get to the lavatory in time.

Escape avoidance

There were some people who avoided thinking about any circumstances that might force them to move. They regarded their decision to stay put as permanent, firmly stating they say they would never move, no matter what happened: 'the only time they'll be taking me out of here is in a box'. They were able to justify this decision to stay put as realistic because they were getting plentiful, freely available help from partners, family, friends and neighbours. However, some of the people who were heavily reliant upon the support of others to stay in their current homes, were effectively protected from having to think about potential problems. They ran the risk that an event such as an illness, or someone moving away or dying might shake the foundations of their support network: they were forced to start contemplating how they would cope without as much help. Although Mrs Ashley was compelled to move due to illness, previously she had begun to realize that she could not always rely on her friends for help.

> *I*: If you'd had a choice (*after becoming very ill*), is there anything different you'd have chosen?
> *Mrs Ashley*: I'd have stayed at home. But then, I'd have had to think again, a bit later on. My person that worked for me had a husband that cleaned my car and took me out in the evening if I wanted to go out, to be dropped somewhere. He was getting older and you can't replace that kind of friends. So, that is really an advantage of being here. (Mrs Ashley, 90 years, widow, residential home, London)

Although recognizing that they may be forced to move, nevertheless, some people were reactive rather than proactive, deliberately preferring to wait until a crisis developed. This may be due to a personality trait, or as in the example below, from a learned helplessness (Seligman, 1975) where people have experienced a lack of control over life events.

> Yes, really it's almost an impossible thing (to plan for the future) isn't it? You don't know what's going to happen tomorrow. We're not planners. Things, I think tend to happen. I think that until four or five years ago (*diagnosed with cancer*) I felt I was in control of things, you know I decided what would happen. As I'm getting older it tends to be now that I'm in a situation where I react to things happening. I think, as you get older that you somehow lose that ability to be in control of a situation. You are more at the stage where you are in position A and you are assailed by various sides, be it health, be it finance, be it whatever and you react to what happens. (Mrs Belcher, married, own house, Lancashire)

Distancing

Although people sometimes choose to distance themselves from having actively to think through a problem situation, they may still be incubating the idea, with the result that they are more receptive to the possibility of moving and alert to opportunities that may present themselves. Whilst they might reject the notion of 'planning ahead', they had nonetheless clearly given their future housing requirements some thought, even if, rather dismissively, they may refer having only 'vague ideas' or 'having it in the back of their minds'. What they appeared to be doing was perhaps better described as a style of mental rehearsal, flicking through possible future scenarios and contemplating what they might do in these circumstances. Even when people say their moves 'just happened', usually it is still possible to identify a series of events and circumstances that led up to their moves.

> Well, it wasn't really a decision, it sort of happened. Because after I retired, my mother and father lived with me and my mother was very ill, she was in hospital for a while and she died. And there seemed to be no point in father and I continuing to live in Hest Bank because we didn't like it particularly. And it was very stormy. We had had the window blown in a few times, it put us off a bit. So I had a former colleague who lived in Abbeystead ... and she said, 'There's a nice house here coming empty', but she said, 'You won't want it, it will be too big for you'. And we had an afternoon riding around and came here and that was it. I knew I didn't want to be anywhere else. (Miss Barrow, 89 years, single, own home, Lancashire)

In this incubation period, in which people are aware that their current housing may not be suitable in the future, they may end up having to 'create a crisis' to develop enough momentum to make a decision. They talk of events escalating, finding that they go along with them because they have been given an opportunity that they should seize. Mrs Panton in a passage already quoted, exemplifies this approach:

> I've always believed that if you are in the right place, the time to move comes at the right time. ... And I looked out of the windows and something said, 'This is right', and I always know when it's right, I don't have any hesitation.

Problem and emotion-focused

Seeking social support

Social support has been defined as information from others that one is loved and cared for, esteemed and valued and part of a network of communication and mutual obligation (Cobb, 1976, in Stroebe, 2000). Social support can be both problem and emotion-focused because people can provide various kinds of support: emotional, instrumental, informational and appraisal.

Emotional support

There were examples of older people actively participating in helping us conduct our research, whilst struggling with their own housing dilemmas, who used the research advisory panels as a source of emotional support or as a means of giving support to others, based on their own experience:

> It was at a meeting I was to attend that I met the person from Age Concern, Newport, who gave me your 'Housing Decisions in Old Age' pamphlet. I thought I could identify with this project because of my own experience. If I were able to make a suggestion, it would be for a counselling for this situation (of being alone and having to move in later life). With the extended family now gone, there are many in this situation. Only last week, I met a lady who had made a very bad mistake concerning her move and was very lonely and since then she has had a stroke. (*Written Housing stories* – Mrs Weldon, 78 years, widow, Isle of Wight)

Advice and information

Older people reported that they felt more confident when getting information from people who had experienced the housing setting that they were considering: they were more confident relying upon personal contacts and their own experiences of specific housing schemes and residential homes.

> And I must mention one other attraction. At that time a very interesting friend of mine lived in this block, but she has since died. But having feared being in a kind of institutionalized setting it was an increased benefit to have one interesting friend there. And I would say that to almost any older person: 'If you can, try and choose a place where you know somebody who has already lived there and can tell you'. You see this was much better than consulting a family, much better, because she'd lived here for two years. (Mrs Carrick, sheltered housing flat, Lancashire)

> *I*: Was there anything else that you felt made it easier to make a decision?
> *Miss Scott*: No, I knew so many (*residential and nursing homes*) you see. I knew lots of others that I'd visited, visiting people from church and various other ways. So I had a pretty good working knowledge. And when my aunt was in the nursing home, both of the nursing homes that she was in, I used to visit her nearly every day, I used to drive from (place name) and visit her nearly every day. (Miss Scott, 84 years, single, nursing home, London)

Some were able to use their contacts to hear about possible vacancies.

> But I was getting a bit creaky because I've had my hip replacement in 1991, so my garden efforts were a bit limited, so I had to have quite a bit of help in the garden. Then I heard, I asked one or two of my very good close friends here if it would be a good idea if I got myself onto the waiting list, and they said, 'Yes that's the thing to do'. And then it was that the opportunity for this one cropped up. I heard about it through a very close, life long friend of mine, who used to live down south near us, who is on the committee here. And she told me that someone wanted to move out of here into the big house, but she didn't know when. (Mr Farley, widower, residential home, Lancashire)

In any case, personal contacts were highly valued sources of advice and information, much more so than government or private sector sources.

Instrumental support

Practical support was seen as essential and highly valued, with many people effectively supported by family and friends. The extract below however, reveals the kinds of practical support that is wanted and just how desperate older people can feel if there is no one to help.

> I have been trying to obtain rented sheltered accommodation for some time, but have not been offered anywhere suitable. I need practical support and I do not have any family and never had the time or opportunity to make friends, due to working long hours, not to earn overtime payment, as I never got this, but to keep my job. Also, always having one or more older people to care for, or help or to care for. I have not got good neighbours, and have no-one to whom I can turn, if I need even a small kindness. Social services cannot provide me with help. I was told I could only have the type of support I need if I were to move to sheltered accommodation.
>
> I need someone to help me to buy food if I am ill, and to change light bulbs, open child-proof bottles and do very small handy jobs. I have had a lot of torment from neighbours and want to feel more safe and secure. The places I have seen have nearly all been full of mould and dampness, not at all healthy. Also, they had such small rooms, almost cupboards so I would not be able to take my books and I would be very bored and lonely. They did not have anywhere to dry washing and some had no room for a washing machine or a fridge/freezer. I could not live like that.
>
> It seems that now, the wardens have been stopped from giving small amounts of help to the occupants of the sheltered accommodation and now the residents are complaining that they get no help, when that is the reason they moved to sheltered accommodation. I would like to stay in my own home if I could find some one to do the things I need, but it is impossible to find people to do these things. I am willing to pay a reasonable amount. I have to bring strangers off the street, passers-by, to help me and offer unaffordable amounts of money and I have been nobbled and beaten up when bringing in strangers, but when there is no-one else, I have to take a chance. A lot of the sheltered accommodation is of very poor

> standard, not well built, and there are a number of complaints about this type of accommodation. I do hope some thought will be given to providing accommodation which is of a decent standard and to providing support to residents who need some practical help in small ways. (*Written Housing stories* – Miss Eaves, 72 years, single, Cardiff)

Appraisal support

One of the reasons why social support is considered to help individuals deal with stressful situations is that other people can help individuals maintain a positive and stable self-esteem, the positive and negative beliefs and evaluations that they hold about themselves. According to Festinger (1954) the assessment of the validity of one's beliefs about 'reality' and about one's own level of ability frequently depends on processes of social comparison, particularly when objective criteria are missing. In deciding when to move, some people found it helpful to compare their situation and circumstances to others.

Some people found the panel meetings helpful as a way of talking about their own experiences and learning from others. One of the difficulties of housing decisions in old age is that of finding the combination of appropriate places and appropriate people with whom to discuss what is available. As members commented in one panel:

> 'We've got lists, but they are usually out of date and not vetted', in relation to trying to find reliable trades people.
> 'You need to be able to trust the person you are talking to, even if they are a stranger to you.'

In the preceding sections, we have argued that housing decisions are influenced by people's abilities to reflect upon what is important to them, their lifestyle choices and the strategies they use to cope with life's pressures and disruptions. We now move on to discussing how they were able, or in some cases not able, to retain control over their housing decisions.

Control over decision-making

The connection between perceptions of personal control and psychological well-being in the general population has been well established. Concepts like 'locus of control', 'learned helplessness', 'fatalism' and 'mastery' have all connected psychological distress to a person's perceptions of their ability to act on their environment and control forces that affect their lives, thus developing a sense of personal control (see Evan, 1981; Mirowsky and Ross, 1984; Rotter, 1966). This is especially evident in disadvantaged populations and powerless groups, as discussed by Kohn (1972) and Wheaton (1983). Clough (1981) used 'control of life style', 'the extent to which a resident has mastery of her life', as a key variable in analysis of residential homes, arguing that the wording was important because it looked at control from the perspective of the individual resident rather than the more typical focus of research at that time on institutional control (pp. 31–2). Rodin (1986) found that increased control was related to greater happiness and lower mortality in nursing home patients.

The extent to which people involved others in decision-making varied from some people deciding to move and not telling children until their minds were firmly made up, to others having their children make the decision for them. Many people were categorical in claiming they made the decision alone. Sometimes deciding alone was a deliberate choice and other times it was from necessity because there was no one they felt they could have involved. In the sample, it was rare to find professionals involved in the decision-making process.

> *I*: When you started to think about this change, and it sounds like your niece was quite involved, who was making the decisions? Was it something you discussed with your niece or with the rest of your family?
>
> *Mrs Wilkinson*: Oh, I did it entirely. I told my family. I told them what I was thinking. But my niece came everywhere with me, she went with me to sign up to get the keys and she came here to see the flat. But the decision was mine entirely. They were all absolutely delighted. They hated me living alone. (Mrs Wilkinson, 84 years, widowed, sheltered housing, London)

> So I went home and rang the secretary and said, 'Yes, I would like to take that room'. Of course, the family said, 'You've just bought your flat, you don't need to live in an old people's home', all this sort of thing. And I said, 'Well, I'm sorry, it's my decision and I know it's right'. So they left it with me. And of course they are all delighted now that I live here. And that's how I came, and I came because

> I knew it was the right thing to do, and I have never looked back really. (Mrs Pontin, 83 years, widow, residential home, London)

Interestingly, in the next extract, although the person describes her satisfaction in taking the initiative with regards to her future housing, she refers to the decision being 'their' decision (that is the family's) rather than 'hers':

> From my point, I am quite sure that I made the right decision and my family are delighted. Incidentally, they have been saved having to make the inevitable, awful decision on what to do with an elderly geriatric, and possibly disabled, mother who needs looking after. We have all been down that road and it is a very difficult decision to make. I feel I have made their decision for them. (*Written Housing stories* – Mrs Carver, 82 years, widow)

Some, unfortunately, had to learn through bitter experience that allowing others to make the decision for them might not result in a successful move. The first comment is repeated from Chapter 1 (p. 13):

> The biggest mistake made to date was listening to other people telling me what I needed or didn't need. With hindsight, I now realize I should have made up my own mind, in fact, only listened to myself. I am the one who has these afflictions. I am the only one who knows what I can or cannot do. (*Written housing stories* – Mrs Bennett, 70 years, Bristol)
>
> I don't think that I could probably say that I was either for or against the idea of coming to Morecambe in the first place. But I did have my own private feelings that I wasn't that keen on it. If I'd have been given the free and open choice I shouldn't have come. But even at this time, I have to consider that we were getting older and less mobile in many ways and my wife's walking was really bad, so whether or not I had misgivings, it didn't really matter much. I was making things as sensible and reasonable to operate here. (Mr Brown, 90 years, widowed, living in bungalow, Lancashire)

In situations where children, worried about the feasibility of their parents continuing to live alone, intervened and urged their parents to move into more supportive accommodation, their efforts were not always appreciated initially. Frequently, however, the decision to move was eventually regarded by the parents as sensible, even if it took a while to accept and get used to it.

I: She lives in Manchester does she?
Mr Carpenter: No, no, she no, no that's why I'm here, she lives in, well she lives in Lancaster. And when things develop the way they did, she had more sense in her little finger than I had. Nothing was going to stop me. I was going to stay there. But she would say, 'But you'll never work these gardens, Dad, the way you did'. I say, 'Why not?' She says, 'Because you're getting older' and I wouldn't have it, of course. I just wouldn't have it. Then I fell and I broke my ankle and had to go to hospital. And she says, 'That's enough, Dad, you are going into a home'. Well that was the last straw. I says, 'I'm not going into any home,' I says, 'Damn me, it's like going into an airport departure lounge, I'm not going into any home, no, no, no, not me'. But again she prevailed at the end of the day. Unbeknown to me she set about having a look round homes and she went through I don't know how many homes in Lancaster. But this one appealed to her more than any other and she had me come up to see it. And I still wasn't, I wasn't very happy about this set up, I must confess, I was not happy about it. (Mr Carpenter, 85 years, widowed, nursing home, Lancashire)

I didn't decide (decision to move into sheltered accommodation). My daughter decided. I had a cancer operation in Newcastle and got over it very well. I had to go every six months here to have a check-up for 5 years. She got worried whilst I was living alone afterwards and I got so much help that I was all right, but she thought, well, being 300 miles away, it was a long way. So she came and looked at a flat for me. She looked at this (present sheltered housing flat) and she was quite taken with it, so when she saw me she said, 'I want you to come down and have a look at a room; it is empty and you can have it'. So I thought about it and I thought, 'Well, yes', only because she was living near. Because I was driving my car until the day I came, so I had to sell it. I didn't let them have it till the day I came. That's how I got here. (Mrs Alton, 90 years, widowed, residential home, London)

Well I couldn't look after myself so my daughter looked round and she liked this so she asked me if I'd like to go in and I said, 'Well yes, I've got to'. I says, 'It's not a case of, "Do I want to go?" I've just got to go'. She said they went all over the place and they looked all around and she said, 'Mother this is the only one that we liked' so she said, 'I'm sure you'll like it'. So that's that. Well I had to. My daughter she said, 'Well mother we can't have you' she said, 'We haven't got the room'. But she said, 'That's up to you, we don't want to force you'. But, well

> I thought 'Well, I've got to do something and I can't do it on my own'. So she said, 'We'll take you to see it'. So they came in a car and we went to see it here. And she said, 'Now what do you think of it?' I said, 'Well they all seem nice people and that'. There's one or two that aren't, but you mustn't take no notice. (Mrs Hilton, 100 years, widowed, residential home, London)

Mr Elridge's story provides an excellent insight into the way children can be involved in the decision-making process. At the invitation of the interviewee, his daughter was present during the interview.

Mr Elridge's story

I: Can you start by telling me about why you came to be living here?
Mr Eldridge: My wife and I were under the daughter's eagle eye for at least ten years, but we had no option. The daughter said, 'Up sticks'.
I: Right, where were you living before you lived here?
Mr Eldridge: Well, we were living in Manchester. We were townies really, and my son-in-law, a forester and he was living in Wales, which was very pretty so we sold up, we had a caravan; we lived in a caravan for about six months. But it depends on the kids really.
I: So your daughter wanted you to move here to be near her?
Mr Eldridge: My son-in-law's firm collapsed, he got a job with a new firm as a forester, chief forester and they had to move up here, followed the job, you see, we're not short of cash. I'm not short. So we had a mass run up into this area and we found this going.
I: Sorry, just to make sure I'm clear, you moved twice to be near your daughter?
Mr Eldridge: Yes, we moved twice.
I: And why did you choose this place in particular?
Mr Eldridge: Well, the job was here; well it was in the next village.
I: So how did you decide on this house?
Mr Eldridge: That was my idea wasn't it?
Daughter: Do you want Dad's perception, or do you want the truth? (Laughter)
I: I'd like to hear what you're both saying.
Daughter: That matters as much anyway, doesn't it? Dad's perceptions.
I: I'd like to know if you have a discrepancy, but I am here to interview you (looks at interviewee).
Mr Eldridge: I could see the gleam in son-in-law's eye. We saw this one and we had a look at one in the next village.
Daughter: I'm biting my tongue.

Mr Eldridge: Which was a bit far away from the job, because the job was up Kirkby Lonsdale way. Am I telling stories? (looks to daughter)
Daughter: Yes.
Mr Eldridge: Right, you say.
Daughter: The truth of the matter is we needed somewhere; we had three children at home.
MrEldridge: Yeah, yeah ...
Daughter: And we wanted you and mum to come with us. When we first decided, when (son-in-law's) job ...
Mr Eldridge: I said that ...
Daughter: No, no, but the problem is simply finding a house big enough.
Mr Eldridge: That's right.
Daughter: It wasn't a matter of choice. It was finding somewhere that would accommodate everyone, mum was backing off, she said, 'I don't really want to move', but she was terminally ill. So we were saying, 'I'm sorry, you need to go, you need to be with us'. We needed to all be together somehow. And there was no way we could have managed with half of us there and them up here. So we just had to find somewhere that was big enough for everyone, didn't we?
Mr Eldridge: (Laughter) You see what I mean?
Daughter: Yeah, it wasn't a matter of choosing. We had to find somewhere big enough to accommodate everyone. And this was ideal, because it was a downstairs flat and it was originally two flats, it was holiday accommodation, wasn't it? So we bought this because mum and dad could stay downstairs with their own bathroom and everything else and we could have the house upstairs. That's why we chose it.
Mr Eldridge: So, you see, it's not the choice of the old folks, is it? I remember once when I was as young as this one. We were on holiday and what's it and there was an old couple with a daughter and son-in-law and to all intents and purposes they were having a good time and I got talking to the old people and their cry was, 'Well we go where we're taken'. And for a long time past, that struck me as being non-cooperative. They were told, 'Well you can't go where you want to go, old people, you've got to go where you're taken'.
Daughter: It's the truth, isn't it? You have to go where there's transport to get you about, really.
Mr Eldridge: If the old folks were on their own, they'd go nowhere, ever. Housing is a minor problem, it's people that matter.

I: So, the decision was taken that you needed to be together. Were there any other options that you considered?
Mr Eldridge: No, not really.
Daughter: Well, at the time we moved, mum had only, she'd been told she only had six months. And the option would have been to leave my husband up here working and us all stay there.
Mr Eldridge: Oh, no, no, no, well that could never have happened.
Daughter: And she herself didn't want that. What she wanted was to see you sorted, with the family. (Mr Eldridge, 79 years, widowed, living in daughter's home, Lancashire)

As one can see, despite the light-hearted banter, there is a deeper underlying issue that older people 'go where they're taken', indeed that people such as Mr Eldridge feel that they have no real choice in housing if they become reliant on their family. The family needs of their children seem to outweigh their own, and compromise becomes necessary.

Family doctors and consultants can be instrumental in persuading older people to move into supported or specialist accommodation, although within the interview sample, the level of professional influence described below was unusual:

> I'm not only concerned about the future, but also now! My husband of 43 years died three years ago and my doctor wants me to move into sheltered accommodation because of my poor health. (*Written housing stories* – Mrs Dent, Surrey)

> We have too many stairs. Everywhere there are stairs. After my last knee operation I said to the consultant, 'What can I do to protect my knee?' And he said, 'Move into a bungalow' so that is one of the things I would have to have in mind, to be on a level surface. But it would have to be, I don't know, an older house I wouldn't want to live on an estate or anything like that. (Mrs Harrison, 64 years, married, own house, London)

Alternatively, some doctors actually had a particular sheltered housing scheme in mind because of the appropriate level of support offered.

I: So when you say it was decided, did you make the decision, or if not, who made the decision?
Mrs Silver: Well, no, it was taken out of my hands. I was too far gone, shall we say, and it was discussed with the doctor, and the doctor said,

> 'Well, of all the homes, I would like her to go to (very sheltered housing scheme). She'd be well looked after', and so of course it was discussed, and my daughters came here and everything was looked into, and so I came here, as I say, for two weeks and then six weeks and then it was decided I had to stay here.
> *I*: So just to get the picture then, before you came here you were quite capable to look after yourself and with your husband?
> *Mrs Silver*: Up to being ill, up to being ill and then suddenly I just went. It took me all my time. My husband's brother was with us for a week before and it used to take me all my time to make the meal and to carry it and then clear up, and I used to be utterly exhausted. And I just thought I was tired or I was not well, I never thought of anything so serious. (Mrs Silver, widowed, very sheltered housing, Lancashire)

The following extract from an interview shows how the task of persuading someone to consider specialist accommodation is hampered by the perceptions and misperceptions they hold about such accommodation.

> *I*: Perhaps your own experiences will help other people make informed choices about housing decisions?
> *Mrs Dennis*: I hope it would, you know, because I knew nothing about any of this sort of thing (sheltered housing). When the doctor said to me, 'We think we'd like you to go to sheltered accommodation', I was horrified, I really was. I said, 'Oh, I'm not ready for that yet'. I don't know what my version of that really was. I'd never sort of talked about going into sheltered accommodation, never wanted to. I was happy where I was.

Later ...

> *I*: So, whilst you were a bit upset about the security aspect at home, you hadn't planned to move?
> *Mrs Dennis*: Oh no, I was going back. I was going back to work and the doctor said, 'You're 80 and you're still going to work?' I said, 'Yes, but it's a very easy job'. It wasn't hard work and so he said, 'No, I don't think you will be'. He said, 'Where did you live?' and I told him and he said, 'You're not walking up and down those hills?' because that road is quite steep, 'We'll have you moved, we'll take you to sheltered accommodation'. I said, 'I'm not ready for that yet'.
> *I*: So what did that conjure up in your mind, 'sheltered accommodation'?

> *Mrs Dennis*: Nursing home, in bed all day, sat by your bed, that's what I thought, when you're in hospital and you get out of bed and you sit in your dressing gown. That's what I thought. I had no idea. (Mrs Dennis, 82 years, sheltered flat, Lancashire)

Perceptions such as these, that sheltered housing, for example, is the same as a 'nursing home', are not at all uncommon. In particular, many of the respondents equated sheltered housing with residential and nursing homes, and so with a life stage at variance with how they saw themselves. There is in that sense a double misperception, both of the range of options available and the differences between sheltered housing, extra care housing, residential and nursing homes.

The ability to exercise control over housing decisions is in practice subtle and complex. When decisions are taken in a crisis, there is less likelihood that they fully incorporate the wishes of older people and more likely that they necessitate considerable compromises to which people become resigned.

Conclusion

In this chapter we have been searching for the key elements that make a housing decision one that someone can feel comfortable with. In one of the final research meetings when looking at whether the emerging findings were 'True for us?', the panel members, the older people who had been interviewers and the research staff all took part in some creative writing. One of the interviewers commented later that the event had been a revelation to him because it was only at that stage that he began to understand the experiences of the people he had been interviewing. 'You have to examine your own experience', he said; 'You can't interpret what others are saying until you can interpret your own'.

It is that reflective and reflexive quality which seems to be central to making decisions that people are happy with: to have a sense of self-understanding and awareness of needs and goals in life generally, and the way housing and location fit into those aspirations. This 'looking glass self' has been seen in the respondents' decision-making. Some have examined their experiences in terms of elements of *personality* which need to be satisfied, while others have prioritized more focused *lifestyle* aspirations. Others used problem-focused *coping strategies* to overcome the specific difficulties which they face. Lastly, there were those for whom control, and perceptions of *self-control,* in the decision-making process were of immense importance. Some had found themselves in positions where they felt the decision was not theirs, and this tended to result in a sense of fatalism. Many others took specific steps to ensure they, and they alone, made their decisions, and this had proved to be a strategy with which they were satisfied.

8
Preferences in Living Arrangements

What people want in their housing

In earlier chapters we have argued that, in seeking to understand the reasons people have for moving or staying, account has to be taken of:

personality and decision-making style;
events that change a steady state;
the suitability of the house; and
the potential to improve their circumstances by moving.

In this chapter we focus on one aspect of housing decisions: what people want from their housing, often shown by what they look for in a house to which they might move. What people look for and value in their housing is a part of the process of decision-making; nevertheless, although artificial, in this chapter we attempt to isolate housing preferences.

Within that context, when thinking about what is most suitable for houses in later life, people want housing that allows (or preferably enables) them to live as they want *and* that is easier to manage as they grow older. The creation of a home is for the purpose of establishing a place where you can live well. People want to be fulfilled rather than diminished by the place where they live. Referring to houses in this way does personify them in that a building is described as a place that 'allows' or 'enables'. Yet, it seems a useful way to portray the interaction between person and building. The way in which space is constructed is discussed in the final chapter.

One way to represent what elders want from housing is to list essentials, such as electric sockets that are convenient in relation to height and location. Such detail is important, and we draw on it throughout this chapter. However, the detail can mask factors that are intangible.

We think that what people want is to be able to create a place where they can live well. Tinker writes of 'the desire of older people to be able to live in the way that they want' (1997, p. 110).

An easier house to manage

In considering a move individuals are trying to find a house that offers significant advantages over the place where they are living. They hope for:

- reducing, or at least stabilizing, expenditure;
- an easier house to manage in terms both of the maintenance of the house (e.g. painting and decorating or repairs) and daily running (not so difficult to keep clean or heat); and
- a place where the design and the fitments result in their being able to cope better with their own greater frailty or lack of mobility (perhaps with no stairs, a shower with a seat or plugs at worktop height).

Finances

For some, the first consideration is to release capital, as these comments highlight:

to gain capital when I retired, not having a private pension;
to down size on retirement and release capital; and
to get more spending money – too big a house and garden.

A greater number of people wanted to reduce the running costs of the house. Some interviewees mentioned specifically looking for accommodation that was economical to run and had efficient heating systems. Others wanted to find houses with lower council tax and water rates. More generally, one person noted that, 'The financial cost of owning a house was too much after being widowed.' There were several people who became aware after the death or disability of a partner that the costs of running the house, literally in terms of money and in terms of their effort, had become too great. Some people just stated that they wanted to live more cheaply. One response combined economy in costs with economy of effort: 'We will be looking for a property that is easy to maintain, *low on running costs*.'

Others were very anxious about having moved to accommodation where they had no control over the service charge:

> I have not yet been here three years and in that time the service charge has risen from £56 per month to £81.21 per month. Only last week we

> all (33 bungalows) had a letter saying they had underestimated for 1999, and we have to pay another £143 each. How could they underestimate by such a large amount? This is an awful worry to most of us. In fact, if it goes up much more I shall not be able to afford to live here.

Tasks that could not be managed

'Somewhere more convenient for later years', the phrase one person used, typifies the approach of preparing for easier living: the tasks that needed to be done had become too much; either people did not want to have others coming in to help or they could not find people to do the work. Gardens were often given as the main difficulty for people. Appleton states that some people who look forward to more time for the garden in their early old age, find later that 'this enthusiasm may be waning, especially if only one partner is left to maintain the garden' (2002, p. 30). Heywood and colleagues note that the problems of looking after a garden are one of the factors that push people to move out of their house and that houses without gardens are a factor in attracting people to new homes. Alongside this are those who have a 'deep attachment to home or garden' and for whom this is a reason to stay where they are (2002, pp. 86, 88). In our study, someone stated:

> The bungalow itself suited my needs perfectly. But when I became unable to garden myself, every summer was *the same struggle to get gardening help*. As for what we are looking for, well, we are still unsure apart from being *smaller, easily maintained gardens*. (our italics)

Some did not want a garden, commenting that 'the land is better utilized as allotments or other surrounding space which can be used more flexibly over time'.

Many found that the management of the house or garden became impossible when one of a couple was unable to do the part of the work that they had done: 'My wife died leaving me alone in a bungalow with a large garden to maintain at the age of 85.'

Of course illness created similar problems. One man in a meeting of one of the panels to review our findings talked about the reality of the situation with which he was faced. In earlier meetings he had described the wish of himself and his wife to stay in the large house which had been their home for 50 years. They wanted to be able to have children and grandchildren to stay, hosting large family events. He wanted space for his large piano. He knew the work his wife did in the house but explained how he had not appreciated the reality of what that meant until she became ill and he struggled to look after her and keep the house going. He found this a sharp lesson in thinking about whether they could stay where they were. Such problems were compounded for those who wrote of repeated ill health, for example, problems with heart and sight, no near relatives and friends too old to help.

Housing designed to help manage disability and reduced mobility

People often had precise reasons for their move. Mrs Goodman set out her reasons for the overriding importance of a place that was warm:

> I made the decision because I need to be here because this room is always warm because of the arthritis and everything. So because of that I need this one. It is always warm for us. As old people we can't live in a cold place. If we go out we can't put the heating on all the day long, but here the storage heating is adjusted according to our body's needs. (Mrs Goodman, sheltered housing, London)

In the *Housing Decisions in Old Age* study, coping with stairs was the single biggest problem people found with managing in their current homes. Although half the questionnaire respondents said they did not have problems with stairs, over 20 per cent of those who had a short flight of stairs and 33 per cent who had a long flight of stairs were having difficulty managing the stairs. Two per cent had installed a stair lift to cope with their own or their partner's difficulties with stairs. A large number (30 per cent) already lived in accommodation with no stairs.

We got more detailed information from interviewees. Some used stairs deliberately as a form of exercise, even if they had other means. One person explained her system:

> Yes, I have got a stair lift and I use that part of the time, first thing in the morning and in the evening mainly because I've got angina and I like to use it in the morning and give my heart a chance to get going. And then I walk up and downstairs all day because I need the exercise and, as I can't walk out, I use the stairs as exercise, well I do walk out but not very much. And then in the evening I have it on again because I do a lot of sitting around if I'm on my own. And it's not good for me to get up from sitting down for two hours and then sort of going upstairs, it sort of gives me angina symptoms. So yes I've got all that sorted out. (Mrs Foxton, widow, London)

Some people had decided that, although managing the stairs was very difficult, they would stay where they were until things got worse. However, the clear picture is that when people had decided to move to easier to manage housing, nearly everybody chose places without stairs, ground floor accommodation or flats with lifts. One person talked about her former house:

> So it was ideal when I was working but when I retired in 1986, then I said to myself I want somewhere where there are no stairs. ... So I think the main thing for coming to where I live now was the fact that I knew quite a lot of people up here and secondly the flat I found was very convenient, no stairs, just two steps down from the main road. Very, very easy for me to manage on my own, things like cleaning. (Mr Spencer, 74 years, London)

Others stressed, similarly, that they wanted ground floor accommodation:

> I had an acre of garden to cope with and I had got rheumatism badly in my knees, couldn't get up the stairs. And I had the chance of a good ground floor flat here, that's why we moved. (Mrs Talbot, 83 years, widow, sheltered housing, London)
>
> I had a first floor maisonette and my wife was taken ill and after a number of falls and other problems, the doctor advised that we obtain ground floor accommodation, which was not easy. I didn't want to go into a block because of high maintenance costs which, if anything happened to me, my wife who was not working would be faced with. So we decided to sell the maisonette and look for ground floor accommodation. We were very fortunate to obtain a little bungalow where we are living now after quite a few problems. (Mr Page, 87 years, married, bungalow, London)

Stairlifts were seen as a means of managing in people's current housing: nobody spoke of looking for a house with a stairlift. There was a downside to stairlifts: some disliked modifications that made them feel like living in an old people's home. They worried also about annual maintenance charges and whether the re-sale value of the house was reduced.

Some people who were more mobile were prepared to move to a house with stairs, but with some provisos. For example, one person said that her main criterion was to have a toilet downstairs because she needed 'to go to the toilet an awful lot'.

Kitchens and bathrooms

There is no doubt that people want housing that is well designed so that their management of daily living arrangements is made as easy as possible. Many were highly critical of arrangements even in places that had been specially designed for older people. Kitchens and bathrooms, rooms at the hub of household activities, were of particular importance in terms of layout and practical arrangements.

This conversation is typical and illustrates a frequently encountered problem of being unable to open kitchen windows:

> *Mrs Bracken*: It's the planning, I think. You see these planners, when they sit down to go to work, they've got one thing in mind: they've got to produce more planning in as small area as much as possible. ... They've got to be all one bedroom, two bedroom whatever the case may be, they've got to fit in with the requirements. ... I mean the kitchen window I have to get a walking stick to open it.
> *I*: Do you really?
> *Mrs Bracken*: But I did with our other, I had to stand on a chair and ours were new windows. ... But I have noticed in that particular part, that what you're talking about is this, in every kitchen that is laid out and planned out, you have to stretch over the taps to reach, to open or shut the windows. It's not funny as you get older really, not very good on balancing. ...
>
> At the club we were talking about getting up steps and things and I said, 'Well I don't know I can remember, I could get up on a chair and bounce up onto the draining board and that, but I can't now'. ... Somebody said, 'The trouble is when you get older you lose your bounce', and it's true you do.
> *I*: Yes and you get less confident as well, I'm terrible. (Mrs Bracken, 82 years, married, recently moved to sheltered housing, London)

The interviewee's husband commented that it would be comparatively easy to design gaps between kitchen sinks and worktops to allow someone to get closer to the window. We know of one housing organization that has developed an electric motor to open and close windows over sinks.

However, when older people were consulted and involved in planning, they praised the thought and design. In particular they liked adjustable heights for worktops and shelves that could be pulled out beside a cooker to place hot dishes. An ironing board that tucked inside a drawer was much admired, as were features that were designed for easy use:

> Even the dinner wagon had a ledge on it and a handle to pull it. A lot of dinner wagons don't, they are just smooth, and they are rather dangerous if you're pushing them, but this was set so somebody could push it easily.
>
> There were a number of compromises really that had to be made – for example with the adjustable height kitchen units. They weren't going to be suitable for the tenants, for the future tenants, because they required more storage underneath the units. At least they were there, so everyone who came to visit could see that it was possible to have that kind of thing.

It is worth noting that in a project *A House for Life*, this *Better Government for Older People* group had been involved with architects in planning the adaptations to be made to a local authority house to make it suitable for older people. Once the adaptations were complete, the house was opened up for the general public to view as a show house. Between them, members undertook many activities, visits and talks to develop their thinking and knowledge through experiential learning that between them totalled 530 hours, a topic that is discussed further in Chapter 9.

Bathrooms too often presented major problems. They were likely to be small and so the baths too were often smaller than the norm in length and width, making manoeuvring very difficult.

> I have to turn over and I find it a job. In the other bath, you see, a bigger bath, it was easy to just roll over and kneel and get up from there. (Mrs Lawton, 69 years, widow, recently moved into sheltered housing, London)

The design of handles often caused difficulties. People described not knowing where to put their hands or how to stand; they worried that they could not lift their legs over the handles. Mrs Lawton graphically captured this later in the interview:

> *Mrs Lawton*: Maybe if there was a physiotherapist or someone to come and show one how to use them effectively but I haven't gone into that yet.

> *I*: But literally the way they're positioned, you just don't find them convenient to use?
> *Mrs Lawton*: No I don't find it terribly convenient.
> *I*: Well it can't be a particularly good design then, can it?
> *Mrs Lawton*: No. And she (staff member) said, 'Please don't try it' but she said, 'It can be done that way'. But my knees don't bend you see, they don't bend as much as they should, there is no power. However, we'll get by.

Mrs Carpenter also commented on manoeuvring in baths:

> Yes, there is a bath, but it's getting down as well, and I sort of have to slide down like that. When I go to get up, I can't get up. I like a shower, so we paid for the shower to be put in. (Mrs Carpenter, married, 3-bedroomed house, London)

Housing to assist people in living as they want: space

The amount of space and the way it is divided illustrate well this mix of detail and objective. Many have assumed that as people age they will want, and indeed should have, less space. Most housing for older people has fewer and smaller rooms than is the norm. While it is correct that many older people move to smaller homes, it is equally important to recognize that many reject special housing on the grounds of size. Two comments from our study typify this group:

> I thought, I've seen *Hawthorn Housing* and I thought they were absolutely awful; they were like rabbit hutches. Because you don't necessarily lose your desire for space or taste just because you've got a bit older and infirm. And I thought really, I didn't know who they were aiming at, but as far as I could see and most of my friends who've looked at them for their parents, they were missing the boat, they were missing the market.
>
> We did look at one that's not too far from here and it was a single flat. Well, to be quite honest, I don't think it's let even now. I know that other people have looked at it, but really and truly it was terribly small. I don't know how you could expect even one person to live in it. And it had no built-in wardrobe in the bedroom. To be quite honest the hall was nearly as big as the bedroom. And when I said to the warden, 'Well it hasn't got a built-in wardrobe', she said, 'Well, you can have a wardrobe in the hall.' And honestly it was so tiny that

> I think it's been empty for months. (Mrs Brown, 82 years, married, sheltered housing, London)

There should no longer be any room for doubt that older people want more space than is allocated to housing designed for elders. This view is supported by Heywood and colleagues arguing that older people's housing should embrace 'a focus on the importance of space rather than a concentration only on issues to do with access' (2002, p. 167). Appleton has a similar emphasis:

> For many the activities and social patterns of old age require at least as much space as life styles in early parts of the life cycle. (2002, p. 25)

The space is required for numerous different purposes:

storage and display;
enough rooms to have people to stay, pursue interests, sometimes separately from a partner, and to separate cooking, eating and sitting;
to get around the house, when people are less sure on their feet, have walking aids or wheel chairs;
to manage household tasks.

The following comments from different people reflect their views on why they want sufficient space:

> My husband and I get on very well but I think it's partly because we've got space because we don't like doing the same things. If I've got the television on and he wants to read a book, he has got a room he can go and read a book in, and vice versa. If he wants to watch the cricket and I want to read a book, then we can do that. We've got enough space so that I have a work room and he has a little office, so that's good as well.
>
> I need an extra bedroom – so if I'm ill my daughter can stay some weekends and sometimes I stay over with her, which gives me a little change.
>
> I had to have a place big enough for somebody to look after me, to have someone to live in. Sometimes my boys stayed overnight if I was frightened.
>
> The character doesn't change just because you get older. I think most people will tell you there's not enough storage. That's always

the thing when architects design things, they make nice spaces but they never ever appear to think what people want to put into them. They don't like cupboards anyway, because cupboards intrude onto their nice spaces.

Two bedrooms. At this age we like to be near enough to know our partner is safe but also able to read in the middle of the night if we cannot sleep.

Several interviewees mentioned the need for sufficient circulation space, recognizing that, when less agile, it becomes *more difficult* to manoeuvre in small spaces:

> But also the house was acceptable because it was big enough to be able to walk round with a Zimmer frame for mum and she'd got an area where she could walk up and down with a Zimmer frame. (Miss Nightingale, single, living in own house, London)
>
> I use crutches and that, you see, when I take my leg off, so I need quite a bit of walking round space, so we have shirts hanging on one door, trousers on the other and the wardrobe jam-packed. (Mrs Dunstable, 72 years, married, sheltered housing, London)

Once again this chimes with others (Appleton, 2002; Askham *et al.*, 1999). The argument becomes even more persuasive if the focus shifts from the detail of what space is wanted, to the purpose of the space. The phrases 'space to …' or 'space for …' extend the discussion into this area. So people will state, for example, that they want a number of rooms, certain size of rooms or, more abstractly, sufficient space. However, frequently an explanation is added: space to be able to have people to stay or to do things separately from their partners.

It is essential that there is consideration of how people live, rather than assumptions of what is needed. In a fascinating study Percival (2002) looks at the use and meaning of domestic space in the lives of older people. Space becomes used to create and manage daily routines, to fulfil responsibilities and to allow personal reflection.

When describing a special place in her house one panellist in our study spoke of 'a huge lounge/diner where everything happened'.

> I don't know if it's the room; whether it adapts itself to my needs or me to it; there's a window; I'm on my own; comfortable; have the TV; there's company when people come; it all works out.

In her former house she had had a huge kitchen/diner which served the same function. The special place had changed with a different house, but the functions continued. Percival has examples of people who eat their meals while watching TV to manage their loneliness. However, there were others whose continuity of lifestyle had been interrupted because they could not both eat and cook in the kitchen. It seems that, in the main, people do not want to cook, eat and live in the same room. If there is no separate dining room, the preference is for a kitchen that is large enough to eat in as well as cook. The alternative, as above, is a large living room but some people did not like having guests to eat there. Some did not have visitors for meals because the place was not suitable. ...'Older people may feel not only the loss of space but also of a personally meaningful role' (Percival, 2002, p. 734). One woman said: 'I feel I've robbed my children and grandchildren of something' (2002, p. 734). Percival continues by claiming that such a home is seen as 'a non-family-friendly environment, with negative implications for the older person's morale and sense of self determination'.

A second theme in Percival's discussion is doing housework, a topic that could be extended to other daily living tasks such as looking after yourself. In effect, the layout, the organization of space, may make doing such tasks easier or more difficult. Percival asserts that the activities themselves contribute to people's sense of control in their lives.

Third, he comments that, 'Responsibilities are highly significant in family interactions' (2002, p. 747). A building may diminish people's capacities to maintain or assume responsibilities. The final point of relevance here is that space is wanted not only for storage but also for setting out possessions, a function that allows personal reflection. If there is no room to display precious items, the playback of memories is less likely. That is not to say that the more mundane lack of storage space is unimportant. It seems incredible that so much special housing has little storage, and that what there is, being low down or high up, may be very difficult to access.

We have focused on space both because older people state that it is important and because it neatly demonstrates the importance of understanding how people construct their lives. Space either allows or inhibits the management of daily living. In the process, it supports or undermines attempts to create the sort of lifestyle that is wanted. In one of the final panel meetings someone commented that too often the space that existed was broken up into too many small areas: she found this restrictive. Similarly, someone wrote in:

> More space is surely the greatest need but there is a plethora of little houses in my chosen area with adequate total space chopped up into small rooms, upstairs and down.

The point relates to the way people live: in later life people will spend more of the day in the places they create to give comfort, security and interest from observing changing scenes or people, as well as to be a base with important items around *and* a place to entertain friends. It is harder in small spaces to construct the place as wanted.

A place that allows or assists people to live as they want

In the five final meetings with research participants, set up to test out our findings, we decided to look further at what people wanted to create in and from their homes. As referred to in Chapter 1, creative writing was used as a medium to encourage reflection. Some individual responses follow in which individuals write about places in their homes that are special:

> It's a lounge that looks out on the road; I knocked down a wall from a small study; a huge lounge, I like the size, the decorations, wood and fittings that I put up; a huge picture window and another small window. I'm home when I am there with features of our family around me.
>
> *An armchair* – it's in the living room, lots of things within range – a cupboard with books; drinks; a table with paper work; the telephone; the chair can tip and I can sleep; it's comfortable; there's a view on to the street and the garden.
> *My summerhouse* – it's in the garden; I can read, watch nature; I'm away from stress.
> *A living room* – there are family pictures around; I feel secure; I don't like change; I like to know where everything is in the dark; I have a fear of going into a home.
> *An armchair in the living room* – it looks out on the street and on greenery; I can see trees through another window and see the changes; books are nearby; there are holiday mementoes; a TV and radio; everything's centralized; there's a fire and it's cosy. ('True for us?' Individual comments at final panel meetings)

In the passages above individuals write about particular rooms or pieces of furniture that capture elements that they value. It is worth noting that, by way of contrast, someone who had been living in a sheltered housing scheme for ten years could not think of the place where she was as home: home for her was India – the garden, hens running round, memories.

On a broader scale, the way that the house is perceived, indeed the meaning that it has, is related to its place. Not surprisingly, the practicalities of living are uppermost in many elders' minds as they reflect on the reasons they chose particular places:

ease of access to shops and services – people were aware that they were becoming less mobile, had more difficulty with walking, and might not be able to continue driving; they took account also of transport systems, though several people reported moving to areas with adequate transport, only to find that services changed;

safety – there were many references to changes that had occurred in the area where they lived; often young people's behaviour was seen as the problem, but in some areas there was a feeling of having to be on one's guard as to what was said because neighbours might report your comments; people related such feelings to places where drug trafficking took place.

In the questionnaire we asked people to list factors that had been, or would, be important to them in moving house:

46 per cent said they would (or did) give considerable weight to the idea of moving to be nearer particular people;
28 per cent strongly agreed that they would consider moving to be nearer particular services;
23 per cent strongly agreed that they would (or did) consider moving to be in an area they knew well; and
18 per cent of people gave some or a lot of thought to moving back to their roots.

Various comments from interviews and panel meetings illustrate the points:

> ... Added to that, the neighbourhood was hilly, and with age I found walking more difficult and I knew my time left as a driver was getting short. There was only one shop for food on the estate. Over that length of time the neighbourhood had radically altered in character. All these reasons made me look for a flat.

One person commented that it was 'difficult to get out – no place to go'. Several added the reason why location was important, for example, wanting 'to be near son (husband died after moved on retirement)', 'to be close and transport easier' and 'to be nearer to brother in law'. Wanting to have easy accessibility to family members combines different aspects. In part this is consequent on the growing importance of family as elders recognize that their time left before death is diminishing. Wanting to be close to people who are special to you, and help to locate your place in the world, is of growing importance for many. However, there is a combination of pleasure and social activity on the one hand and, on the other, a feeling that family members may be the most reliable for advice and, possibly, direct care. 'Being accessible' relates both to wanting to be able to travel to family members with as little trouble as possible, and also to hoping to reduce their travel time in visiting.

In the following sections we look at some of the activities of daily living that occupy people's thoughts as they think about where to live. Two extracts capture the centrality of access to shops:

> Another thing is I don't drive, my husband drives. So where we were, we were in a good position, I must admit, for the shops and things but then when my husband was ill, I had to just take my trolley and go to the shops on my own. ... As I got older, it got more worrying. (Mrs Carroll, married, living in own house, London)
>
> I realized that my wife couldn't be supported in that particular environment – it was a mile and a half from the shops. There were local shops but you couldn't get a wheelchair in them, so that was no good and, therefore, we decided to look at possible alternatives. (Mr Slater, 62 years, living in sheltered flat, Lancashire)

Over two-thirds of black and minority ethnic community elders rated nearness to shops as important compared to 47 per cent of white householders in a study by Boyo (2001) (quoted in Appleton, 2002, p. 18).

Transport illustrates the interconnectedness between one factor and another.

> The neighbourhood was hilly, and with age I found walking more difficult and knew my time left as a driver was getting short. (*Written housing stories* – Miss Membler, 87 years, single)
>
> I want to move into something nearer the town centre because, although my present flat is only a mile away, the bus service is threatened and the service finishes at six p.m. anyway. (*Written housing stories* – Mrs Casson)

> The bus service is terrible and I have to rely on dial-a-ride, who are wonderful. We only have one shop in the village and that is closing next week. I would like to keep my independence as long as possible, but it gets harder every day. Our post office suddenly closed for 10 weeks earlier in the year, it's reopened, but for how long? To get to the next post office would cost me £2.50 by dial a ride, which is a lot out of a small pension. (*Written housing stories* – Mrs Sellers, Birmingham)

Thus, the place of transport in the organizing of where to live is linked to the person's own capacity to walk or drive, the services that are available locally and the quality of public transport. If anyone of these changes, then the viability of others may be called into question.

Phillips and colleagues (1999) compare the poor transport system in the UK to Scandinavian countries where the role of transport services in community care is recognized (quoted in Appleton, 2002, p. 18).

A factor that loomed large in many people's assessments was access to health services. We did not check on the frequency with which they actually went to health services, but there is no doubt that it was judged to be a very important consideration. People's own perceptions of their state of health and of the risk of ill health come into play. A further aspect relates to the actions to be taken to counter the risk: some would want to ensure that they have access to a phone or an emergency call system with a further safety feature being someone who calls round regularly; others, as the comments below show, want to have easy access to a hospital:

> It's no place to live if you're elderly because the hospital in the town is not very big. It's very good, but it's not very big and it only has visiting consultants, and Winchester is a wretched journey away. So I would never advise anybody to move here. In fact one couple are moving out now, a couple up the road, their mother is in a home miles away. They can still afford to run a car but I don't know for how long, so I don't think country living is the thing for people unless you've got a decent hospital within decent travelling distance. (Miss Harbottle, 79 years, single, own flat, London)

> Well, we're planning to stay here. We've got to be realistic now in life. We are 20 minutes away from all the hospitals and you're 2 or 3 minutes away from the shops. All right, 30 years ago when you're young and you're virile, you go somewhere and you make your life there and you become part of that community. (Mr Thomas, 64 years, married, living in own home, London)

Several people mentioned that they wanted to be able to get to their temple:

> *I*: So you had to come here. Why did you choose this part of East Ham?
> *Mrs Joshi*: Because the temples are close by and I don't need to depend on anyone to take me about. (Mrs Joshi, 73 years, widowed, sheltered housing, London)

> *I*: Why did you come here?
> *Mrs Bhatti*: I have got only one daughter; she lives in the same road on the top. And the temples are here, and the shops are here. And community, most of the community, we mix with people in our community; they live here in this area. And we can buy our traditional food and everything is near this place. It's not too far from doctor's surgery and the post office. (Mrs Bhatti, widowed, sheltered flat, London)

> *I*: So then you were not happy in Newcastle, you wanted to be near ...?
> *Mrs Ghazi*: The children, everybody goes to work and school so I am lonely there. So I need to live here because most of our Asian people were here. And I go to the temple where I can meet them. (Mrs Ghazi, sheltered housing flat, London)

Once again the importance of seeing the links between features is illustrated from the accounts: people want to be near the temple but also note that they are lonely and want to meet others, that they do not want to be dependent on others and that other key facilities are near the temple. Christianity was mentioned less directly, though some did comment in links to other topics as when, in talking about a retirement village, a reference was made to good neighbours who took a couple to church which, they said, had always been important to them.

The ability of local neighbourhoods to support people's shopping, social, leisure, work, health, religious, cultural and transport needs was frequently given as a reason for staying or moving. New locations were assessed upon these criteria. Not surprisingly, specialist housing schemes were rejected if they were built in the wrong place and people anticipated being isolated from their communities.

Another consideration in determining whether and where to move was the difficulty to which several respondents referred of meeting new people.

> There are so many people who retire early and come to live in Lanford, so it's not like a town or a village right in the country where people never change. They are always changing here and retiring and coming full of enthusiasm wanting to serve the community, so it's a good place to live. And there are lots of groups going on – music groups and study groups and literary groups, so I just belong to those. (Mrs Panton, 83 years, widow, living in own home, Cumbria)
>
> Stay here, yes. This is an ideal place for me because if my daughter wanted to go somewhere else, I don't go. I won't go with her. I don't like to live with her. I want to be free and I feel you know, our community, they are good for me, they are good friends. I've got the feeling it's okay. (Mrs Baxter, widowed, sheltered flat, London)

Many people attributed staying put to feeling daunted about the prospect of building up a new social life. They questioned the wisdom of uprooting and leaving friends and links to the community behind.

> *Mrs Knight*: Well when you move like we did .... It is very difficult to make new friends and I would always bear this in mind for wherever anybody wants to go, because for the first five years I was here I was literally miserable.
> *I*: Were you really?
> *Mrs Knight*: (Yes), because where I lived in Hampstead, because we'd been born and bred there, we knew literally hundreds of people and there was always people dropping in and out the house and you know, it was lovely. They'd come in for coffee or tea or lunch, it didn't matter what, but there was always people around. Whereas, living up here, I never saw a soul. And I go out and walk down the road and I see someone and would say, 'Good morning, good afternoon' and their nose would go up in the air, and ... they were very snooty when we first came to this road.
> *I*: So having said all that, do you plan to stay here for the rest of your life?
> *Mrs Knight*: Well if I moved now and if anything happened because my husband is older than I am, I would be absolutely lost. So I think it is so essential to hold on to the things that you know. I would love to stay here for the rest of my life; whether I will be able to afford to if I was left on my own, is another matter. (Mrs Knight, 71 years, married, living in own house, London)

Some described choosing a particular village because they already knew people there. Another person spoke about how they had weighed up the attractions of a new place against the fact that they knew people in the locality where they lived:

> Well, we did think about moving at one stage. I suppose about ten, twelve years ago when we came back from Africa and ... we looked at places. Well, we thought, move out to Thame first of all and had a look out there; couldn't actually see anywhere we liked but we only went once to check. And then we got quite interested in this 50+ Leisure Group in the village and that made us think, well, perhaps we'll stay rather more locally because we couldn't have got to it from Thame and the people of course, that's a plus. (Mr Parker, married, living in own home for 33 years, London)

It is the likelihood of getting to know people that attracts some to housing communities. One person describes this, in a passage which has been quoted earlier:

> For me, I'm somebody who has to be part of the community as well and so I have got friends all around me and because we're all in the same boat here, we all look after each other. (Mrs Raynes, widowed, 1-bedroomed cottage, retirement village, London)

Others emphasized continuing with work type activities. Lord Johnston stated that, in choosing where to live, he chose a housing company with a good reputation, wanted a place that was not too small and was close to Westminster to continue with his work in the Lords.

> I intend to continue to play some part while I can do so usefully. I have already reduced the amount of work I do, as a general rule I confine myself to asking questions rather than to take part in debates, because if you take part in debates you've got to stay there for the whole of the time. (Lord Johnston, 93 years, widowed, residential home, London)

Another said:

> I was still a member of the Esher Wednesday Circle and I'm still included in anything that they want me to do. And at that time I was driving and I was still able to go over to Esher to see my friends and

> whatever. And I still carried on doing my voluntary work from there. I used to bring people down to have their feet done here or run them to the doctor's. (Mrs Walton, 88 years, widowed, living in own house, London)

A different dimension is illustrated by those who stress that they want to move to a place that is pleasanter. Some commented on changes to their area, often because the place had got busier and traffic levels had changed dramatically.

> Now we wish to go to a less noisy area, with less pollution. We are only five minutes walk away from the M32. Our road has 92 houses and it is a no through road. (*Written housing stories* – Mr Danby, 56 years, Bristol)

> My only thought was I was going to give my mother a better environment, better breathing conditions, because it's much cleaner here, in the last years of her life. We've got more trees and it's much cleaner. The atmosphere is cleaner. (Miss Noakes, single, living in own house, London)

> Yes, yes, because it would be nice to just say on a nice evening, 'Oh let's go for a walk'. ... If you wanted to walk to a place where you could actually get among the trees and grass and whatnot, it was at least half an hour's walk to get there and half an hour to come back. (Mrs Murphy, 60 years, married, living at home, Lancashire)

Many people suffered from poor sound insulation. This is a common criticism in flats but becomes an even greater problem in old age, as stressed to us by older people themselves since the volume levels on radios and televisions often are turned to maximum. Someone else commented that her bed was set against a wall, with her neighbour's kitchen being on the other side. She was disturbed every morning soon after five by the water pipes as the neighbour made her morning cup of tea. High levels of sound insulation are essential. One person said that the prime reason for her move was to get 'quietness especially at night'.

A further negative aspect that led some older people to move was the problems which had arisen when new neighbours moved into an area, for instance, students and families with young, disruptive and noisy children.

> It used to be very friendly, mostly elderly people who had lived here all their lives. In the 18 years we have lived here, a lot of residents

> have died or moved into care and their houses have been taken over and converted into flats or bed sits for students. One or two houses have been purchased by the council to accommodate problem families and some have been bought by ordinary householders. The thing is every household seems to have two or three cars. ... the parking is terrible. Many of the vehicles have to wait to get past or turn around, which means we have to put up with their music blasting out while they wait. Another thing that has spoilt our road is when people knock down their front garden wall so they can get another car in and nearly always leaving walls and gates down and unfinished. (*Written housing stories* – Mr Danby)

Someone else cited a similar set of problems and ended:

> And eventually I said to my husband, 'We don't have to live here and sit here worrying about it all the time, let's get out.' (Mrs Mason, 60 years, married, living at home, Lancashire)

Mrs Mason recognized an opportunity in later life to change their lives.

The place of housing

In setting out what people stated they want in their housing it is hard to get the balance right as to the importance of housing in people's lives. Some say that they can make their home wherever they live, so it might be assumed that less attention should be paid to housing quality. Others assert that elders have the same rights to high quality housing as everybody else. Both statements are clearly correct. In the same way, while some emphasized the importance of storage space, others were making the point that moving house to smaller accommodation provided an opportunity to clear out clutter. They stressed that they wanted to be free of many of their possessions: 'material things have less importance; photos of shared experiences matter more'.

There is no paradox. People can, and do, create their own living environment in whatever circumstances they find themselves. Yet in later life, people should have high quality housing, though they will probably want less overall space than they have had through their lives. Specialist housing with one bedroom is likely to be as unwanted as much of the sheltered housing of the 1950s and 1960s.

We have argued throughout this book that what people are looking for in their housing cannot be understood solely by trusting a formula of needs and wants to produce a solution. However, it has been apparent from earlier chapters that, in making housing decisions, people trade-off one quality for another. Insofar as people are conscious of such transactions, this is a rational process. To take an obvious example, it may not be possible to live within walking distance of shops and services *and* to live in the country. In addition there may be some aspects in which there are competing benefits within a category. Privacy provides an example. People may be very clear that they want privacy in that they want to be able to live their lives as they wish without being accountable to others for what they do. Yet at the same time they may be very anxious about becoming ill or having a fall and being left alone with nobody knowing. Thus a decision may be made which in effect acknowledges that the concern not to be left alone when in trouble overrides the delight in being private. Arrangements are made for a housing warden to phone in every morning at a set time to check the person is all right. The construction of home and lifestyle is negotiated between the house and the people who live there, indeed between the house in place and the people in place. This is illustrated by another extract from Miss Davidson's experience:

> *Miss Davidson*: I think the majority of people would like to come into something which is equivalent to what they've been used to only with the facility to be under supervision should you need it. We have our bell pulls here, the warden rings us every morning, but I mean, you don't feel you're under great supervision or anything, you just know that if you need help, it is there. I wasn't going to have the bell pull at first. I thought I'm far too young to. ...
> *I*: Could I ask how old you are?
> *Miss Davidson*: I'm 60. But then I had a terrible chest infection in the spring and I was really poorly and the warden insisted on phoning me every day and she said I really think when you're better, I'd like you to stay on and let me ring every morning. She said just because you think you're young and fit doesn't mean to say that suddenly one night you're really poorly, and you could be lying there for 2 or 3 days and nobody would notice. And I thought well, yes, it's probably a good thing, so I do get a ring most mornings unless I'm going out in which case I ring them. But apart from that it's just like living in a normal house. (Miss Davidson, 60 years, single, living in council owned, sheltered housing 2-bedroomed bungalow, Cumbria)

Housing, in the words of one panellist, 'is probably the most important contribution to quality of life'. People (with their current resources of health and money, and their feelings of self-worth) interact with a house (which may be more or less easy to manage, and more or less fulfilling as a place to live). It is impossible to predict whether a particular house will allow someone to be as they want to in their old age, although there are many key aspects that can be noted. We are clear, as were the respondents, that in later life citizens have the right to share in improvements in housing. As we argue in the next chapter, elders should have the best housing that society can provide: they spend most time in the house and they need most support.

9
Theory, Policy and Practice

Introduction

The focus of this book has been on analysing and interpreting people's housing decisions. In this final chapter we attempt to locate the themes that emerged in their wider worlds. Thus we look first at the implications of what we have written for theorizing in different areas, in particular decision-making, ageing and research methods. Finally we examine the relationship between our conclusions and policy and practice, especially in relationship to housing but also more widely with reference to the involvement of older people as citizens.

Decision-making

People arrive at their decisions in different ways. Some people's decision-making styles could be described as being more rational, in the sense that they use a more systematic, structured approach that is based upon extensive research, long-term planning and calculating whether alternatives meet a set of desirable criteria. Others do things differently. They may use their understanding of themselves, emotions and intuition to reach their decision. People also use different types of beliefs and knowledge systems, explicit and tacit to make their own judgements as to the risks and benefits of certain courses of action. To date, the research in this field has had a strong bias towards quantitative research and has tried to predict 'moving behaviour' by using theoretical models of decision-making. Quantitative research is good at providing explanations of behaviour at a societal level and capturing what is explicit. We believe the dominance of the rational decision-making models and styles is a reflection of the type of research methods that have been employed in the past.

The qualitative approach used in the *Housing Decisions in Old Age* study, treats the person as a whole and enables explanations of housing decision-making to be located at both a personal and intra-personal level. In their housing stories, people have described both the concrete, tangible and external reasons why they decided to move or stay where they are, as well has the impact of their inner lives, their thoughts, beliefs and feelings, on their decisions and choice of actions. Collecting personal housing stories reveals how differences in personality, age, sex, biography and culture influence both the circumstances in which the story tellers find themselves and how they respond.

To understand housing decisions in later life it is necessary to capture the structural influences on people's wider worlds, such as the neighbourhood facilities and transport systems, the availability of suitable ordinary and specialist housing in their locality and the influence of their own personal circumstances – their health and mobility, together with the help they may be getting from others. At a deeper level though, elders are reflecting upon how best to use the limited time they have left, aware that their reduced physical capacities and energy means daily chores take longer and so time to enjoy oneself is proportionally less. For some, moving to specialist accommodation is a deliberate strategy to free up time.

Ageing

Decisions about where to live in old age are of immense importance for older people themselves and for housing policy. Later we consider both what would help people make the decisions that will fit best with how they want to live and the relevance of our findings for house building. In this section we review older people's experiences of their recent housing journeys: how do they fit with theories of growing older.

Studying their own housing situations leads people to reflect on their lives. Housing is a central aspect of people's lives, playing a part in how they perceive themselves, how they present themselves to others and how they manage daily living. There are dangers in isolating housing in that it may be ascribed too great or too small a part in people's assessment of their quality of life. It is clear that it is both basic and integral to people's quality of life. However, as well as being a factor that can be isolated and studied, it is intricately tied in with other core parts of people's lives.

Some people argued to us that they could make their home anywhere: the essence to creating a home came from within them. It is not exactly that, like snails, they carried their home around with them. Rather, they stated that home could be created wherever they happened to be and in whatever circumstances. Such people were emphasizing the importance of their internal worlds; the key resources were theirs. The picture is one of people with a solid internal core that allows them to adapt to the circumstances in which they find themselves. Yet another picture is equally valid. In the second scenario people cannot live how they want to because of either the constraints of the housing or of their social environment. Both assertions offer insight to the experience of growing older.

We have some evidence of the experiences of ageing of people from ethnic minorities. Nationally, 35 per cent of owner occupied property in the worst condition is occupied by ethnic minority people, and this echoed a recurrent theme from some of the ethnic minority panellists in the capital who had encountered many problems in maintaining their homes as they grew older. While they did not report different problems than their white counterparts the problems were more prevalent because they were living in older housing and in an urban area.

There were many different experiences recounted of housing journeys. Some spoke of moves from the Indian sub-continent to Africa, only to find that they had to flee from an oppressive regime to England. Others had migrated themselves from the Caribbean, or had parents who had done so.

When discussing their roots, panellists from all ethnic groups had very different conceptions. Two had grown up in children's homes and found the notion of roots alien. One man had fled as a child from Nazi Germany. A woman wanted to return to 'her roots', which turned out to be a place where her mother's family had lived, but she had never lived herself. Clearly roots have strong emotional and symbolic resonances. The 'myth of return' was expressed within two of the panels. One of them, in Lambeth being comprised wholly of African Caribbean and African Asian elders, addressed this issue on a number of occasions, recounting stories of their own or an acquaintance's vision of a better life 'back home'. That vision did not come to fruition in the way that the dreamer might have hoped for and there were incidences of men working to build houses back in the Caribbean only to die before completion.

The same contrast between dream and reality was reported by a white panellist who had thought about going back to live in her childhood village. She discovered that the people and the place had changed. Presumably she had changed herself, though she did not mention this.

Quality of life is a term that has largely superseded earlier concentration on *successful ageing*. In part the move has followed from recognition of the obvious problems associated with trying to define as well as to assess success in ageing. People's assessment of their quality of life seems to tie down more concretely factors such as health and capacity to manage daily living that are seen as essentials.

In similar vein, we can (and do) set out the aspects of housing that people state have a bearing on their ability to manage well in the places where they live. Yet there is a danger that highlighting, for example, the importance of space downplays the place of people's inner lives. The richest accounts we have been given of people's housing decisions, interplay the concrete with the intangible: in these people look at themselves, at who and what they are, and they recognize that, at best, their housing is one part of the creation and re-creation of their lives. In other words, it is those theories that highlight older age as a time of review of one's life that locate best the stories people told us.

This is the life story school, for example, of Erikson who sees the struggle in old age as being between ego integrity and despair. The core question underlying such approaches is the sense that people make of their lives. This in turn links with older people's judgement of their self-worth. Housing decisions cannot be understood unless they are placed in the greater stories that people tell themselves and others of their lives.

It is this relationship between how individuals see themselves and the available housing that explains both what people want in their housing and how they manage what they get. It explains too the relationship between possessions and sense of self. Thus, some people contended that certain things were essential to their construction of themselves: without them they would be reduced as people. Others asserted they could be themselves, or construct their homes, anywhere. The reality is that this juxtaposition simplifies and distorts.

In the creative writing exercises at the end of the study, participants were asked to think about special places and what made the places special. In the discussion that followed, some spoke of the few possessions that were really important, often photographs of earlier events in their lives or of people who were special. What seems to be happening is that there are stages in both conceptions of housing and of ageing, though of course these do not follow a chronological line. Hepworth (2000) writes similarly:

> The essence of symbolic interactionism is the conceptualization of human life as a process. The idea of process implies human experience

> as a kind of reflexive ebb and flow rather than an undeviating movement in one direction. (p. 17)

Discussing the processes of interaction through which identity is created in everyday life, he notes that 'humans manipulate symbols to make sense of the world'.

Another central aspect of the stories that people told us of their housing is that they reinforce the image of ageing as part of the 'journey of life'. For us, this is a counter weight to those, often younger people, who see ageing only in terms of decline. The analogy of the journey is important because it recognizes that the journey continues. Of course, the reality of ageing is also that of physical decline, shown in the regularity with which people reported the problems created for them by health and disability. Our point is that people's housing experiences cannot be understood if the dominant perspective is that of managing decline. What is happening is that people are managing change, both in terms of reduced physical capacity, but also changes in their worlds (the people in their lives) *and* in their views of their own lives: what is important and what they want to do.

And so, as happens throughout life, people are both holding on to parts of their former lives and moving on: this process is individual. Forty years ago there were debates as to whether people aged best by maintaining activities to replace work or be letting go of involvements, disengagement as the theory was termed. By and large activity theory became accepted as the most appropriate model but it is apparent from the participants in this study that 'disengagement' describes well some who want to let go of what could be termed the 'tyranny of possessions'. Once again, rather than validating a position in an 'either/or' debate, the experiences as described to us much better fit a complex model in which both activity and disengagement are seen as important in the changes in old age.

Finally, in trying to locate the study in theories of ageing, we look at the decisions themselves. We did not set out to rate people's decisions either on their own scale or that of others. However, as we have struggled to unravel the data, we have become aware that there is one style that seems particularly effective. Some people seem to have discovered a style in which, first, they examine through reflection what are the aspects in living that are most important to them; they follow this in making their housing decision by trying to ensure that, whether they move or stay, the place provides what is most important. This seems to be a combination of wisdom about self and a style of working out what are the critical factors.

The reflective process is an important aspect but can be achieved in many different ways: internal conversations, writing or talking to other people. As a consequence of the relationship of their decision to fundamentals, and of the process of searching for the best solution, such people are probably also most likely to be content with the decision that they have made. However, even when they have discovered this style, sometimes, the changes that can happen (to oneself, one's partner, one's family, finances or neighbourhood) are so unpredictable that no housing solution in old age is future proof.

Research methods

In social science research there are debates as to the best ways to capture experience. Consideration is given to objectives and processes: what is to be looked at? who is to do it? how is to be undertaken? in what ways is the information to be disseminated? At the heart of debates are concerns about the authority of those who are being studied: what say do they have in objectives and processes. Participatory research, discussed in Chapter 2, has challenged earlier assumptions.

The approach discussed in this book has a bearing on discussions of methods. First, we have described attempts to involve older people in research in different ways. Second, we have reflected both on the problems that arose from research which involved older people but had not been initiated by them and on the ways that we tried to manage the consequences. Third, we have discussed types of involvement, arguing that it is essential to see research as a complex activity that should not be analysed solely on a uni-dimensional scale of control.

Thus, in considering the nature and reality of participation by those who have been seen as research subjects, it is useful to consider where control has lain. However, there has been a tendency, sometimes accompanied by rhetoric, to see greater control as intrinsically better than less control. This has led to tables which equate 'minimal control' with 'tokenism'. Our experience is that involvement in research leads some people to demand greater participation and more responsibility but that others want to limit their contribution. In our study, the people on the panels who were invited to six or seven meetings through the life of the project were less likely to want greater responsibility than did those who had been educated as researchers. There are two possible explanations for this: first, the panellists were older, had greater problems with health and mobility, and some were already involved actively in their communities; second, the fact that the interviewers had been through an extensive educational course meant that their understanding of the activity of research, their skills and their knowledge had all been heightened.

Three points stand out. The first is that people may have chosen to limit their involvement but yet be highly satisfied and see their contribution to an important debate as real. There is no doubt that there are great concerns about the time and energy given to consultation, which many suspect leads to little in terms of outcome. So what is felt to be real involvement is important. The second factor is the importance of education in helping develop skills and understanding. Many people need to develop new skills to maximize their participation in new areas. This leads to the third key factor: much writing on participation relates to people who are service users. Older people who become involved in research may or may not be users of social care services. It is essential to recognize that older people have skills and experience as well as being likely to have limited experience of research. The implications of these points are developed in the next section.

Practical ways to improve housing decisions

We have argued that one of the reasons why some people appear to make more successful housing decisions than others is because they have spent time reflecting upon their lifestyle, personality and what is important to them. This also means reflecting upon what their future housing needs might be, recognizing their own and their partner's resources and limitations, together with their ability to develop coping strategies. Some people are naturally reflective, but others might gain from gentle promptings to start this process as a way of mentally preparing for possible future eventualities.

Decision-making styles

During the course of the research, we found or heard about approaches that were thought helpful to developing a 'looking glass' style of reflection and getting information about available housing.

A structured approach A structured approach has some strengths in either prompting people to start thinking about housing needs in later life or, if they have started to think about moving, helping them to focus upon some of the key issues. For instance, the questionnaire respondents commented that completing the questionnaire had been a useful exercise:

> I live in a potential flood area and this throws up additional considerations with future housing. Filling in this form turns one's mind

> more seriously to the whole subject, perhaps for the first time. So I find the task helpful in that respect.

As mentioned in Chapter 4, the *Housing Options for Older People* (HOOP) tool is already available to help with this, but at the time of writing, needs to be more widely publicized amongst housing professionals and older people themselves.

The HOOP tools were intended to provide 'a thorough and holistic assessment of a person's current housing situation' (Heywood *et al.*, 1999, p. 1). Heywood and Galvin[1] are concerned that the original version is already too time-consuming to complete. It is worth noting that the HOOP tools do not take people from thinking about their possible housing move to a decision, nor should they be expected to. They should be seen as a resource, akin to a discussion giving advice on key features identified in a *Which* type magazine survey. Whilst beneficial, such materials do not fully represent a person's housing dilemmas, nor can they remove the heartache of decision-making.

The nine categories of the HOOP tools were originally taken from research identifying reasons frequently given as to why people had been forced to move house; they appear to be heavily influenced by 'push' factors. In a recent developmental workshop, participants suggested that the HOOP tools could be promoted as an approach or concept. It was recognized that thought could be given to the way the HOOP tools are used in practice, for example, as a mechanism for initiating discussions and thinking about the issues.

An educational approach Such an approach can help people think through issues and make informed decisions. Here were a couple of suggestions from older people on what might make easier choosing whether, when and where to move:

Courses or seminars specifically arranged for older people who need to make housing decisions – these to contain all options possible (including for/against same).

Advertise Local Authority lectures on options available to support older people.

Experiential activities including events such as the creative writing used in this research, creative activities, visits and talks provide opportunities to explore the issues from a personal perspective as well as get additional information.

A creative approach Creative writing did help people reflect on their own situation in a deeper way, identifying emotions, meanings and aspirations. As we mentioned in Chapter 7, one of the interviewers found the 'creative writing activity' a revelation to him because it enabled him to begin to understand the experiences of the people he had been interviewing. 'You have to examine your own experience', he said; 'You can't interpret what others are saying until you can interpret your own.'

A sharing approach The panel meetings provided a supportive setting for hearing others' housing experiences; members learnt from each other about how they coped and what they had learnt. Becoming aware of the impact on people's lives of sudden events or changes in circumstances, facilities and services spurred participants to think about their own situations.

A whole life approach By this we mean stepping back and taking a wider view of a person's life, rather than focusing solely upon their immediate housing needs. It demands looking at people's own experiences, rather than imposing a 'professional' or restrictive view of their situation. This might mean taking a biographical perspective, or talking about their deeper emotions or aspirations. When designing housing, some architects say they would ask their clients about their lives: what do you do? how do you live? They would then find ways to get people to consider options for their housing, perhaps by showing them photographs to try to find out how they want their life to be. Only then would they relate their skills in architecture to providing housing solutions. The parallel is clear in terms of providing advice on housing.

Getting information

Seeing the housing decisions of older people as like any other decision as a consumer fails to recognize that good information is only one aspect of decision-making: as well as having too little information, people may have too much; further, they may not have the energy to act as the model consumer, collecting information and checking options. Nevertheless, the quality of information available is one important aspect of decision-making.

Older people frequently told us that they did not know about key pieces of information that might have influenced them to choose differently. Thus, one person who had been involved in a *Better Government for Older People* group said that she liked living where she was but that she would have made a different decision had she known then what she had discovered as a member of a group involved in looking at housing. Written information is said to be hard to come by, to keep up-to-date or, if available, to understand.

But information by itself is not enough, and often needs to be amplified through discussion with other people. Many respondents had found individuals, usually family members, who had helped them make a decision. However, there were many others who commented that neither estate agents nor local authority housing departments were able to offer impartial advice. Several did not want to rely solely on their children either because they knew that they had an interest in the decision or because they did not want the decision taken out of their hands.

Many people commented that participating in the research had led them or others to whom they talked to re-think their housing options. People did not know what information they needed, until they began to think about what might happen or to face up to problems that emerged as a result of changes, whether to their own or their partner's physical state, or to the wider community in which they lived. Consequently people will not all need the same information, in the same way, nor will they all want it at the same point in their lives.

Of course there are well established factors that impede the value of information: impenetrable language and unnecessary jargon magnify difficulties for those people whose grasp of written or spoken English may be poor.

The comments that follow set out a number of views from the research participants on sources of information and advice.

Source of information

> Advice about housing for elderly and younger people with health problems/disabilities should be made available through estate agents and local authority/social services. A symbol could be awarded to agencies supporting the scheme.
>
> I feel that information regarding housing for the elderly is in lots of cases not easily available. Advice is usually not given until the need arises. There appears to always be a shortage of the right type of places.

Housing information

> More awareness of forthcoming suitable housing in and around own neighbourhood. More knowledge in layman's terms of available grants & entitlements, support systems i.e. decorating, handymen, gardening, shopping assistance, cleaning windows, etc.

I suspect that the decision often has to be made without sufficient warning, therefore it would be helpful to have immediately available in leaflet form up to date facts on benefits, allowances, availability of housing in the neighbourhood, costs of accommodation.

Impartial/expert housing information

- getting clear unbiased picture of future housing options;
- impartial information given on request to a local telephone number might be helpful so that one could classify one's thoughts without pressure from family or friends.

Advice to others

- Don't leave it too late to decide. Do it while you are able.
- Be proactive and make plans and decisions before the ability to act is lost. Seek family advice, at least for moral support.

Going 'Barefoot'

Some people referred to the lack of another person to talk to and offer advice. When asked where they would like to go for advice, there was a preference for non-statutory sources of help, with predictable brand names such as Age Concern and Citizens Advice scoring highly. It also highlights the importance people attach to being advised by organizations that do not have, or appear to have a direct interest in the outcome of a discussion about housing needs.

Many of those we talked with expressed a wish to talk through their options in a considered way with someone whose knowledge, experience and disposition made them easy to talk to, and sympathetic. As people described an ideal adviser it sometimes seemed that they were outlining someone like themselves. The research method of using elders as interviewers may well have had an impact on the interviewees, as well as alerting other members of the research team to the potential resource of employing older people to advise others.

The idea of 'barefoot' housing advisors, people who are not paid professionals but who have interest, knowledge and a commitment to helping others, has worked in other settings and is worthy of being considered. Such an approach could use some of the untapped energy which we and others have observed as we conducted the research. It would create an opportunity for purposeful engagement in community life for people who may have little interest in serving on committees, forums or review groups but who do have a sense of duty and a desire to be an active citizen.

Careful thought and planning would be needed in order to find ways of 'licensing' and supporting people who wished to fulfil this role. When discussing the arrangements for interviews, the panellists in this research had pointed out the scope for exploitation of potentially vulnerable people. Vetting would need to be sensitively carried out, given that the starting point, or a key criterion for appointing someone to the role, would be self-selection. Users of such a service could be protected, at least in part, by the organizational location of advisors, possibly in voluntary organizations. Such bodies have the capacity to select, support with information updates, supervise, organize their appointments and provide quality control, in the same way as they do with other services such as befriending.

Advisers will need knowledge and skills. In spite of the fact that there is no single approach to decision-making that will solve all potential problems, we have no doubt that some styles work better than others. The key components to the more successful decisions are:

- sufficient information about housing options;
- people to talk to who have specialist knowledge, have no direct interests in the outcome, and can help people reflect on the implications of the different choices;
- rehearsal time: time spent imagining what to do, for example, if one's health changed or you could not drive *and* time spent planning the best options;
- being reflective: searching at what is really important to you, taking account of what you know about yourself;
- ensuring that older people are as fully involved as possible.

There is clear evidence also that the most critical housing decisions, those that relate to whether people move from their current home and, in particular, those concerned with a possible move to either specialist housing or a residential home, are more likely than other decisions to be taken (a) in a rush, (b) with very limited information and (c) with less involvement of the older person. Again, there are implications for any adviser.

This approach of involving older people in advising on housing fits well with promoting an active old age, with fostering independence and with encouraging people to take personal responsibility while also contributing to healthier communities in which people feel, and are, more included.

Involvement and participation

It is widely acknowledged that people do not simply want to have things done to them, becoming a client and developing a dependent relationship to an individual or organization providing them with a service. Politicians also express the view that they would like to enjoy a closer relationship with the population they serve with more people participating more fully in civic structures, higher turnouts in polls, and a greater sense of partnership in tackling issues.

Some of those whom we met during the course of the research had been actively involved in their communities in many ways at different points in their lives, but many became involved as a consequence of the research. The experience of the pensioner's movement and of the Better Government programme suggests that there is a blueprint for greater involvement, which might lead to patterns of activity and greater numbers of activists.

But while these moves towards making government locally and nationally more responsive to older people have been one of the more welcome policy developments of the past years, problems persist. Older people may be asked to give their views and to contribute their wisdom to discussion of issues that affect their lives but many doubt whether the energy they give to taking part makes any difference. People seemed weary of being canvassed and consulted; instead, they wanted to see evidence that account had been taken of their views. Their test of the real value of participation is the extent to which things actually change.

Spin-offs from the research project illustrate two ways in which older people might be involved in their communities. Both were unplanned by the research team and followed the initiatives of participants. The first example is of one of the interviewers who trained on the original research methods course and then attended two follow-up courses. He has used his skills to look at the housing available for older people in his locality and has lobbied local councillors.

The second illustration comes from the two groups of older people who were recruited and trained to work as researchers. One group has subsequently developed a cooperative business and undertaken work for a number of organizations, driven on by a mix of motives, including wanting to make a difference and the interest of working in a field that is different from their former careers. They undertake paid work in research, for example, conducting a study into the outcomes of home care services for a local authority. They have advised on the structure of the interview schedule as well carrying out the interviews. As with any organization, they look at their skills and will call on other expertise as needed.

The other group has not developed in the same way as a group, though some of them have been employed as interviewers by a voluntary organization that had secured a contract it did not have the capacity to run without additional labour. One of the London researchers, through the *Better Government for Older People* 'Ruskin Project' initiative, has gained a one-term residential place at Oxford University to conduct his own research project under supervision and pursue an interest arising from his working life in Africa. He describes what he wants to study:

> Much of my working life has been involved with the provision of water to rural people in Africa and, of necessity, contact with non governmental organizations and international bodies. I have seen the practicalities of life for the 'subsistence' farmer and noted aspects that appear to need further study and if possible, to propose some improvements. The developed world's general perception of that farmer, is often of an ill-clad toiler in the sun, with nothing of value to his name. This is a casual view which does not take into account, or value, skills and real wealth. Apparently, he has no disposable income, but may have either a year's supply of food in store, or the seeds and skills to feed himself and family – for the next year – or more. Study should be useful to see what could make him wealthy by developed country standards and whether this would be a good thing for him. Would greater disposable income actually be a benefit to him or not? Assuming that the Western way is Good, what list of priorities can be defined for his development? Development is not taking place at present: there is a need to know why.

Involvement and skill development results in people following routes that they may not have expected or may only have dreamt about.

One ingredient that contributed to these developments was that time and attention from members of the research team were available, at some cost, in one area to support, nurture and tease out ideas. The first projects the group undertook were alongside, or under the auspices of established researchers. Community development, whether focused on a locality or an interest group, does not occur without some kind of stimulation and, at least in the early stages, some continuing support by way of appropriately skilled staff time from public agencies.

Joined-up lives

In the past few years the phrase 'joined-up government' has become more prominent, conveying a sense that government departments are taking a wider view of the impact of their policies, setting out to ensure that plans are more comprehensive, and recognizing the links between the responsibilities and priorities of departments of state.

When individuals confront their housing wants and needs, they balance, for example, location against type of property, and proximity to services against ease of access to their home. Lives are not lived in the compartments of government departments. Feeling safe and secure within and outside one's home is a combination of practical steps, for which the individual may be wholly or partly responsible, coupled with concerns about the wider community to which local and central government, as well as less formal community and neighbourhood organizations, all contribute. These examples are magnified and multiplied in many ways as people take stock of the things that are most important to them.

The challenge for public policy is to respond adequately to the inter connected, sometimes fuzzy links between what people want and need and the ways in which bureaucracies are structured to respond to them. Enabling people to exercise control over their lives for as long as they have the mental capacity to do so, should be a realistic, achievable and measurable goal for public policy. Choice in public services, whether relating to care support or education is now a central platform in government thinking and should be extended to enabling choice in living arrangements in later life. However, as we have argued earlier, it is not good enough to declare that people want independence and choice: involvement and control of lifestyle are other central features of high quality services.

The importance of housing in people's lives

The quality of housing, and people's experience of housing make a tremendous difference to people's lives in old age. This has been well documented and it is not a surprising finding that the respondents emphasized both the difficulty of deciding where to live and how central their house was in their lives. Our research bears evidence to the importance people attach to their houses, and to the ways in which they want to make the place where they live their home. They describe graphically the special places they created, places from which they can look out on a changing world, have access to books, files, television, phone and computer, and where they can relax with friends. The clear implication for housing policy is that older people should share in the general rise in housing standards; they should not always be last in the queue.

What people want from housing

In earlier chapters we have noted what older people stated they want in their housing:

- sufficient space in terms of: number of rooms; size of rooms; circulation space and storage;
- good sound insulation;
- ease of management: design that helps them manage daily living; efficient and not expensive to run;
- to feel safe;
- to enjoy the location.

There are several basic messages for policy and practice. The term 'assistive technology' is used to describe 'traditional' services such as 'wheelchair services and "standard" community equipment' and 'some of the newer electronic assistive technologies of telecare and telehealth' (Audit Commission, 2004, p. 2). There should be no doubt that both traditional and newer technologies have the potential to make major contributions to older people's capacities to manage. However, the term could be adapted more generally for housing as 'assisted living'. It is incredible that, as housing standards have risen significantly in relation to design and equipment, so many older people find that their houses do the opposite of assisting them: too many live in disabling environments.

Houses and equipment should 'assist' first in the basics of daily living in that they make people's lives easier, not harder. So minimum standards for housing for older people should ensure that there are good and adaptable systems for:

using kitchen equipment;
having a bath;
using cupboards and storage;
cleaning;
heating the water and the house;
maintenance; and
moving around in the house, including ease of circulation.

Of course, as many have argued, these should be minimal requirements in all housing, because they are equally relevant for people with disabilities and households with young children. The call for 'lifetime housing' has made the case cogently. However, as the very minimum, housing for older people should be an improvement in terms of ease of management and support for living than the general standard of housing. Too often it is worse.

Second, we make a point that challenges the conventional wisdom on size of housing. There is clear evidence from our study that people want 'sufficient' space. For most this means being able to get around their house easily and having space to mould where they live to how they want to live. A total of three bed and reception rooms would seem the minimum. There is no doubt also that many want a large enough space that it can be adapted to how they want to live. This is particularly important for the main living area where people want to establish their comfort zone, with their essentials around them.

Living alongside people over 60

One of the great uncertainties in planning housing developments is whether older people want to live alongside other older people, whether in small or large communities. Opinions are strongly held on both sides of the argument. A part of the ambivalence would appear to be that people want some of the things that are found in age-specific communities, but are not necessarily wanting to live alongside other older people. Thus, there is a lot of evidence that older people are very concerned about, first, safety and, second, ease and contentment in living where they do. Some described moving from one area because of how life had changed in their community. Others described how guarded they were in whom they spoke to lest it be thought that they were informing on drug pushers in their area.

A specific dimension of the dilemmas with which people are faced relates to children. Some spoke of wanting to be in mixed-age communities and noted how they liked to watch children playing. On the other hand, there was anxiety that older children could seem, or were, threatening.

In the main what is liked about the age exclusive communities is that people feel safe, there is help on hand and a lot of the worries about daily living are removed. In addition, some people said that they liked to be able to go away and not worry about the safety or the management of their house while they were away.

Another central factor in considering such a move was that, typically, it meant moving away from their local community: 'If I could find sheltered housing where I would like to be – LOCATION – local church, library, shops, not isolated – I would move.' Thus, in considering a move to such a housing development, as in every other housing decision, people weighed up the advantages and disadvantages: the aspects that were generally wanted such as quality of housing and accessibility to services did not disappear.

A number of the respondents who did make a move to age exclusive housing were pleasantly surprised by both the quality of the housing and the advantages of the life style. A synopsis from our research sets out the pros and cons:

> The idea of living alongside other elders is less attractive than the reality. Opinions on the acceptability of 'age-related communities' were mixed. Views were divided and strongly held. However, some who disliked the thought of housing provision exclusively for older people changed their views after moving to carefully designed, sensitively managed schemes. Some said they would only move if they felt they would not be cut off from the wider community.
>
> Over a third of questionnaire respondents said they would consider sheltered housing *now*, if it were available in their local neighbourhood, with or without warden or support services. Over half said they would stay put, if they had adaptations and support services. A small number would definitely consider moving into a residential or nursing home now.
>
> Asked what options they would consider if they were living alone and had a disability or long-term illness that severely limited their ability to look after themselves, responses were as follows: the proportion saying they definitely would consider moving into a very sheltered housing scheme with warden and extra support services increased by 29 per cent; those saying they would reluctantly consider moving into residential and nursing homes rose by 12 per cent and 14 per cent respectively. The amount of people who said they would *not* consider very sheltered housing dropped from 39 per cent to 9 per cent. The reasons given for not liking the thought of housing restricted to older people were:
>
> - the housing would not be available in the right location, usually in the area where they lived;
> - the prospect of living only with other older people;
> - community living;
> - living with a lot of people older than them;

- restrictions to the way they wanted to live their lives – shortage of space, rules and regulations, not being suited to the culture and way of life;
- fears of being isolated.

> People's views on living alongside other older people intermix aspects of security, having access to help and company. Of course some parts of this are likely to go hand in hand with others: having availability of support staff is easier in settings where there is a viable number of houses clustered together. (Clough *et al.*, 2003, p. 43)

Involvement with the community life was thought by some people to be an essential to enjoying living in such an environment. Respondents pointed out that some residents did not like the proximity of others. Several noted that they had to resist the blandishments of scheme managers to join in events in which they had no interest. Yet another section clearly enjoyed an active involvement in planning and running events. The positive side of such a move is captured in the following:

> I am pleased and thankful I took this step and I have found living here well above my expectations. I have a feeling of contentment and safety and have contact with people in similar conditions. If I feel in need of solitude, I can stay in my comfortable flat. Otherwise, we have various activities in which we can take part. I find the manager (as she is now called) especially caring and sympathetic and a person to tell of any worries. If ill, she is the first to call a doctor. I know I have been right in coming here and my family do not worry knowing that I am being happy and contented. I would recommend sheltered accommodation to anyone in similar circumstances.

Others would recommend specialist housing because 'it takes the stress away':

> If you live in a house on your own and something happens, you have got to go outside for help. You've either to call a neighbour or you've to get a yellow pages and look for a plumber or an electrician or whatever. And in sheltered housing that is all taken care of and that lifts a great weight off older people.

Finally, many stated that they moved into sheltered housing or a residential home because they were determined not to live with their children. Typically, this was expressed in terms of 'not wanting to be a burden'.

Our earlier discussion of the significance of public and private living space in later life merits further consideration for its policy implications. It seems to be the case that as people age being free of worries related to housing becomes of increasing importance to many older people. In terms of living arrangements there are numbers of people who rate the advantages of living alongside other older people to be greater than the disadvantage: people at a similar life stage, with similar patterns of living where people feel both at ease and safe contrasted with the vibrancy of mixed-age housing, with feelings of being part of ordinary communities. The test for policy-makers seems to be twofold: building age-specific housing in smaller units that are closer to people's current homes; and, second, developing an organizational and community life style that allows people to be involved with their neighbours in the community and with housing managers in ways that they like. Too often, people have felt demands imposed on them.

Help with managing: creating a home; running the home

There are numbers of people who find some aspect of running their home difficult. Getting the balance right in making this statement is difficult. Most people in our study stated that they did not have problems in looking after themselves or managing their homes: they either did these things for themselves or paid for help. Table 9.1 suggests that for questionnaire respondents household maintenance and administration were not particularly problematic. Seven per cent or less indicated having 'a lot' of trouble accomplishing any of the household tasks listed.

Table 9.2 shows the results of asking respondents: 'Apart from what you do yourself, how much help do you get from other people to help you look after your home?' Most of the time, respondents did not receive any help from anyone else. Of those respondents who were married, or living with a partner, 73 per cent said they received frequent help, 30 per cent said the help was regular, but 12 per cent said they only received occasional or emergency help from their spouse or partner.

The fact that most people cope, does nothing to lessen the fact that, in line with other studies, we found that there are many older people who find some parts of managing the home difficult or impossible. These include:

getting repairs done to the house: finding reliable people; paying for the work;

what are described as handy person tasks: changing a light bulb; sorting out fuses that blow; fitting a tap washer;
organizing the home following a move: moving furniture around; hanging pictures;
getting advice on what to do when faced with problems and maintenance; making decisions.

There are two policy implications. The first is further development of the sorts of schemes managed by some voluntary agencies whereby lists are compiled of reliable people to do small jobs and of organizations that give advice on funding. Second, some elders want to have someone to help decide questions such as : 'Should I pay for a new boiler?' or, at an extreme, 'Do I pay for a new roof?'

Table 9.1 Doing/getting help with jobs in home and garden

Tasks	**A lot (e.g. I can't manage alone *and I can't* get/afford the help I need)**	**Some (e.g. I cannot manage alone, *but I can* get/afford the help I need)**	**A little (e.g. I find these jobs difficult and/or expensive so they don't get done)**	**No problems**
External housing maintenance (e.g. roof repairs, window replacement)	39 (7%)	257 (46%)	34 (6%)	191 (34%)
Internal housing maintenance (e.g. decoration, D-I-Y)	35 (6%)	206 (37%)	41 (7%)	254 (45%)
Odd jobs (e.g. changing light bulbs, plugs)	19 (3%)	106 (19%)	12 (2%)	398 (71%)
Garden maintenance	19 (3%)	133 (24%)	29 (5%)	326 (58%)
Heavy housework (e.g. hoovering, spring cleaning)	18 (3%)	117 (21%)	16 (3%)	389 (69%)
Administration, (e.g. paying bills, getting people to carry out servicing and repairs)	14 (3%)	73 (13%)	20 (4%)	427 (76%)

Table 9.2 Source and frequency of help with jobs maintaining home and garden

Help from others	Husband/ wife/partner*	Family	Friends/ neighbours	Volunteers	Home help	Trades people
Frequent (at least once a day)	194 (73%)	23	6	1	10	2
Regular (at least once a week)	30 (11%)	63	22	4	42	16
Occasional (at least once a month)	7 (3%)	88	52	5	7	34
Emergency	3 (1%)	101	103	8	1	82
None	20 (8%)	196	267	402	387	303
Not applicable	297	—	—	—	—	—
Refused	12	92	113	143	116	126
Total	563	563	563	563	563	563

Note: * Percentages excluded respondents, who were single, widowed, separated, that is, not applicable

When discussing our findings, one older person living on his own talked of isolation in decision-making. He described how things go round in your head, an unhelpful sort of 'talking to yourself'. Reliable outsiders play an essential role in helping to think through decisions. Individuals who are on their own having spent a long period in their lives making decisions with their partner may find this particularly difficult but indecisiveness is not confined to individuals on their own: couples may become immobilized. The earlier discussions of decision-making are again relevant, though not of the same magnitude as those on whether and where to move. Similarly, older people as advisers may have a valuable role to play here. We are suggesting that there is an activity that is beyond 'befriending' and different from counselling.

'Towards freedom and beyond domination'

Social gerontologists have shown the ways in which the construction of ageing impacts on the daily lives of older people: change the perceptions that are held of the experiences of ageing, it would be argued, and you would change the treatment, lifestyle and services for older people. Our research is one more strand in weaving the cloth of older people's experience.

Phillipson and Thompson (1997) argue for a critical review of people's own practices: 'a preparedness to reconsider established patterns of practice'. Such a review should 'challenge stereotypes of dependency', 'give people choices', 'focus on self esteem' and 'recognize oppression'. They continue by asserting the importance of partnership as a model of practice: older people should be involved in both the diagnosis and the solutions to problems (pp. 17–18).

The idea of not only understanding but also challenging assumptions is cited by Heywood and colleagues as lying at the heart of critical social policy (2002). They quote from Moody (1988):

> A critical gerontology must also offer a positive idea of humans' development: that is ageing as movement towards freedom beyond domination (autonomy, wisdom, transcendence). Without this emancipatory discourse (i.e. an expanded image of ageing) we have no means to orient ourselves in struggling against current forms of domination. (pp. 32–3)

The fundamentals of challenging current practices and demanding greater involvement of older people are picked up in much official policy. The *National Service Framework* asserts the importance of:

treating older people as individuals with their own needs, circumstances and priorities. All services should respect their dignity and choices – and never make assumptions on the basis of age;

better joined-up working between the various agencies involved with older people.

Better Government for Older People sets out its objectives as follows:

> BGOP exists to ensure that older people are engaged as citizens at all levels of decision-making, and shapes the development of strategies and services for an ageing population. We provide a unique opportunity to share information and learning, develop partnerships, and influence policy and practice. (BGOP, 2002)

Dewar and colleagues in a recent report for the Scottish Executive noted that barriers to further involvement of older people included:

> Negative attitudes towards older people, older people's low expectations of the effectiveness of involvement and a variety of organizational barriers. (2004, p. 1)

They note the need for 'more capacity building opportunities' and stressed that older people wanted to be 'in the know', welcoming 'opportunities for self-development through education and information sessions' and more informal approaches such as buddying and mentoring.

The final piece of evidence we put forward in setting out the current climate is from Wistow and colleagues (2003)

> This evidence suggests that the two most important elements of a strategy for successful ageing should be *control* and *interdependence* in contrast to the more conventional description of *choice* and *independence* as the core public policy goals for people in later life. There is evidence that mortality and morbidity are more strongly related to the experience of control over one's life than the exposure to health risks, *per se*. (p. 4)

They want to see the term 'active ageing' extend beyond physical ageing to 'successful ageing' and 'living well'.

There is no one model for the involvement of older people. Our experience of developing ways to involve older people more fully led us to question our assumptions as to roles in research. This is not to assert that research into matters that affect older people has to be led and undertaken by elders. Rather it is to recognize that there is no preset framework: the more older people are involved in shaping their worlds, the more they will influence and define not only the agenda but also new ways to undertake the work.

One of the aspects of our research project that has most interested others has been the style of involvement of older people. At times this has been frustrating as we have wanted some attention paid to the findings about housing decisions. Nevertheless, it is in this interaction between findings that emerge from experiences of older people and styles of research work, that both new understandings and new ways of working have the potential to develop.

In conclusion we list some events that we have reported on in this study where the detail captures a wider truth:

writing about the special places he had constructed in his house led one man to examine his housing experience;
people who had been trained as researchers wanted greater involvement in research, had more training, and set out to work as researchers;
an individual used his research knowledge to get involved in changing housing policies in his locality;
people demanded more space in housing for older people.

Space for living

Miss Simpson wrote her view of the sort of housing she would like in the future:

Miss Simpson's story
I saw your letter in the local paper as I was cleaning up this morning. I hope this is not too late for you to use. I'm 63 and living in a three bedroom semi, set in an oasis, in a desert of brick paved drives. I would like to stay at home for as long as possible. I am trying to adapt the garden so there's not so much to keep tidy. Not too tidy, the frogs, insects and hedgehogs don't like it too tidy.

I'm on income support but, if I had enough, I'd like a toilet and shower downstairs. As I get weaker, I'd need a home help, but the one thing I'd like would be a buggy. They are a great help to two ladies I know. They can get out on a good day, to shop or visit friends. If I live long enough to have to go in a home, I'd like a home near other people, not out in the country. I'd like a room large enough to be able to not feel I was in a rabbit hutch.

If I could have a flat (I'm single) I would like a separate bathroom. The idea of having to take any visitors into my bedroom appals me. I'd also like to take my cat with me. It's bad enough having to give up my home, without having to murder my poor cat. I'm a person who needs to be by myself at times. In a flat that would be OK, but in a home you have no choice. Why do the people who run homes assume that we all like being treated as if we are in the infants?

I would like exercise classes and talks. There ought to be a room set aside for these things. Art is good, but one man I knew had his brushes and paint taken off him because he splashed on the carpet. If there was a special room it wouldn't matter. When I think of that, euthanasia seems very tempting.

I sometimes wonder if caravan designers could come up with the answer. They seem to be able to pack a lot into a small space. The flats are so small.

Another thing I'd need is that my friends could come, on the bus to visit. Not three miles up a country lane and no buses. Above all, I'd like to be treated like a human being, not as a unit. That's what the NHS call us now, units not people.

The home in *Waiting for God* was a good example, without the manager. When I had to go into hospital or nursing home, I'd like a ward like a hospice where I would be cared for, kept dry, warm and

> fed till my time came. (*Written housing stories,* Miss Simpson, 63 years, Nuneaton)

In this book there have been numerous references to space and now we collect up some of those strands. We try to weave the strands to create a cloth entitled *Living Space*. We imagine the cloth as a picture which is not only beautiful but also subtle and complex so that those who look at it see different patterns and interpret it in different ways.

Putting aside the dream, we are aware of the importance to people of space. In part the demand is for more space: Miss Simpson does not want to live in a rabbit hutch. In part people want larger areas of space so that they can create their own living style within it. Some want better organization, not only in absolute terms of size and shape, but also in the flexibility with which the space can be organized: the smaller the overall space, the more this was judged important.

Miss Simpson provides other clues about the context in which people live and in which they organize their space to create a place to live. First, the location of the house has an impact on how people manage their lives: she wants a place from which she can get out to see other people and to which they can come to visit. Second, she wants good enough facilities to aid her life style: a buggy and a downstairs toilet and shower. The third point she stresses is that there should be rooms for activities such as art classes. Fourth, there is emphasis on how she wants to use space: she does not want to have take people into her bedroom. In different places she highlights the importance of the attitudes of staff, the fifth factor. Finally she asserts that she wants to be treated as a human being and, presumably, thinks that this would happen in a hospice.

There have been other themes discussed in this book which have relevance here:

The idea of home as doing As people live in a place, for example, looking after themselves and the house, socializing and enjoying leisure activities, they mould the building and are moulded by the building.

The notion of home as lived experience In their houses individuals create living spaces, living in that they both communicate to others the lives and lifestyles of the elders who live there and are capable of reflecting back to the elders what is important in their lives. Places reflect a life being lived and so there is space for, and evidence of hobbies, pastimes, interests, passions, routines, studying, entertaining, sharing time with others and spending time alone. In part this is about people's standing in their own eyes and those of others but it is also about countering the ageist view of older people leading narrow, restrictive lives.

Homes as embodied places People leave their imprint upon a place. This is a real physical relationship with a place: the way it physically reflects you, your work and values, and the way it is adapted physically to your physical needs. In a different approach, the feeling of a place is vividly expressed by the idea of ghosts who continue to inhabit the house.

However, space has an impact on living arrangements in other ways. What are the reasons why some places seem heartless and sterile and others are vibrant? The obvious answer is to look at the relationship between the building and the people who live in it, noting the way space is used and experienced, as referred to earlier. There are additional pointers.

The first of these is the construction of the building. One aspect is the way in which architects take their brief. They may take instructions from the developers or future managers of housing schemes, rather than those who are to reside in the building. The participants in this study were dismissive of such practice and demanded involvement. Yet, even if they want to involve older people, architects and planners may not be skilful in eliciting what is the essence of what people want.

The evidence from this study is that most people have limited knowledge of the sorts of buildings and design that are available; they know little of what specialist buildings and lifestyles are like nor do they know of the way their own homes could be adapted. As significant, they are not likely to have worked out what are the most important factors to enable them to live as they want. If the quality of what is built is to match older people's aspirations, architects, designers, developers, and housing managers ought to find ways to help elders describe what really matters. Many will say that they do ask for opinions, and that they know the answers from the places that people choose to live. Such mechanisms do provide some information but it is limited and partial.

In this research project we have begun to capture the perspectives of older people. As discussed earlier, there are ways that can help people to reflect on what is most important to them. One such is to develop the sort of approach that encourages individuals to reflect on their lifestyle and to talk about what they enjoy doing. There is no one way to do this but it can be done by getting people to discuss their daily lives, looking at family photos or writing about special places in the house or special events. Writing seems a particularly powerful way of collecting information on the way places are experienced through all the senses, not just the visual. The means of collecting information may vary; the aim is to build a picture of what matters most in daily living. It is at that stage that an architect can help people to look at the building systems that could be used to produce what is wanted.

The next theme is that places should not just reflect people's age by focusing on handrails, lifts, hoists for baths and floors that are easy to clean. In similar vein Torrington, as we have quoted earlier, writes about the subtleties of daily life: the importance of people experiencing the rhythms of living, day and night, summer and winter. She has a powerful example of the sterility consequent on building corridors (which are spaces between rooms) with no natural light and with lights left on 24 hours a day. The risk of falling may have been minimized. But the result is that the pattern of a day, the rhythm of life, is lost.

There is a link here with the way ageing is theorized and constructed. If later life is seen as space and time that has to be filled, the building becomes a place for people to fill in time between living and dying. Indeed, if later life is seen neither as time for living nor as time to prepare for dying, both the living and the dying may be poor experiences. On the other hand, if life is seen as a journey, then elders are people who have things to do and the house is the place where many of those things are undertaken. It is only if the overriding assumption is of elders as people with lives to live that an appropriate balance can be created between making places on the one hand safe and easy to live in and, on the other, enjoyable and purposeful.

'Just because we are retired and not 100 per cent fit, does not mean we cannot have hobbies. So many people have become vegetated in flats because of no facilities', wrote Mrs Patterson. Living involves doing and for those who have energy for more than maintenance of self and house there are numerous demands for space for activities – sheds, crafts or hobbies rooms, meetings rooms, and places for quiet pursuits (library, reading room, study and computer room). Given that there is unlikely to be sufficient space in a private house for all of these activities, it is interesting to conjecture on ways in which people can have private space in the form of their own flat and access to communal or shared facilities for some events. Some extra care housing and retirement villages attempt to do this and it may be useful to consider the sorts of trade-off that can create houses that are manageable for elders and at the same time ensure the leading of fulfilled lives.

Thus, in conclusion, we emphasize that an essential ingredient in creating good places to live is to understand the lived experience of older people. Without that, architects and planners will design houses to manage physical limitations and frailty rather than to support people in living as they want. There seems little room for doubt that the key to understanding the lived experience is to find ways of talking and listening to what older people have to say.

The *Housing Decisions in Old Age* research was enriched by the participation of older people. Research and housing must be informed by their vision. The findings are presented not as a model of how to undertake research but as one approach, adapted during the life of the project, to involving those who want their views to be heard. It has been a report of a research journey. The hope is that it contributes to understanding the housing journeys of older people. Research can contribute to creating new pictures of people's lives. Such pictures can change the understandings of others. In an exciting way, people may discover new talents and paint their own pictures.

Appendix 1: Interview Schedule

Section 1 Deciding where to live

(Housing Decision Process – last housing move)

1. How long have you been *living at your present address*?

   Prompts: How did you come to be there?
   Where were you living before?
   History of prior house moves – *especially* since retirement?

2. What were the *reasons for deciding to move* to this *area* (county, city, town, village etc.)?

   Prompts: *Were there any other reasons?*
   What was your over-riding reason? Why?

3. Does this area/location *meet your needs* – transport, shops, health services, religious, cultural, social?

   Prompts: Explain how your needs are met/not met
   Is the area safe?
   Do you feel secure here?

4. *Why did you move* to this *place* (house/flat/home) in particular?

   Prompts: What kinds of things made you like the place? Why?
   What kinds of things put you off? Why?
   What kinds of things did or do you think are absolutely essential? Why?

5. Can you tell me *what sort of accommodation it is* and what it consists of?

   Prompts: Prompt for *list of rooms* ... Does it have a bed-sitting room, living room, dining room, other reception rooms, kitchen, bathroom, toilet or cloakroom, bedrooms?
   Do you have to *share* any of these rooms with anyone else? If so, which?
   What *community facilities/rooms* do you have access to?
   Any *services available*?
   Does you have a *garden/open area* to sit in and enjoy?

6. What *other alternatives/options* did you consider?

   Prompts: What were the pros and cons of these options?
   Did you find there was enough suitable property or housing schemes in your chosen area?
   Why was it suitable/not suitable?

7. *Who* was involved in making the decision about where you should live?

   Prompts: Anyone else?
   Can you explain how they were involved?
   Whose opinion was most important?

8. How long had you been *planning* this move?

   Prompts: Can you describe your plans?

9. What had you *ideally* been looking for in your new area and/or housing?

   Prompt for list of *essential* and *desirable* requirements.

10. What *compromises* did you have to make?

    Prompt for *list of factors* which people had to *trade-off* against each other.

11. What kinds of *help, advice or information* did you need (or receive) to make your decision?

    Prompts: Where did (or would) you go to get it?
    Who gave (or would you like to give) this help, advice or information to you?
    Was this information useful? Why?
    What information is needed?
    How would you like to receive such information?

12. What *other pressures or considerations* did you have to weigh up when making your decision?

    Prompts: Finance
    Health (own or husband/wife's health)
    Council restrictions or regulations or others
    Changes to local area (e.g. crime, environment, people).

13. How much did you consider how your (or your husband's/wife's) *needs* (e.g. frailty or ill-health, disability) *might change as you became older*?

    Prompts: What things did you take into account and why?

Section 2 Deciding to stay put

(Housing decision-making process – deciding to stay)

14. Can you explain your *reasons for staying here* (so long)?

    Prompts: Other reasons?
    Perhaps ... People, community, house itself?
    What do you feel is the *over-riding* reason?

15. Were there occasions when *you nearly moved but didn't*?

    Prompts: *Tell me about these. Why didn't you move after all?*

16. What does your *home mean to you*?

17. Do you think if you moved you could *re-create* this in another place?

    Prompts: Why or why not?

18. Do you *plan to stay here* for rest of your life?

    Prompts: *Are there circumstances* in which you might you consider/envisage moving?
    Has anyone else suggested you might need or wish to move?

19. Have you made *any changes to the house* to help you live here?

    Prompt: for a *list of current housing adaptations*, for example, ramps, handrails stairlifts and others.
    Would you *consider housing adaptations in the future*?
    Please explain.

20. Have you *had any help or care* to make living here easier?

    Prompts: garden, house maintenance, domestic help, shopping, personal care (dressing, bathing, meals) and so on.
    Is there anything else that would make it easier to stay here? (e.g. attitudes to new systems/technology – care alarm, internet shopping, gadgets!)

21. What *kinds of information or advice* would make staying here easier?

    Prompts: Care and repair, lists of organizations and people who provide services at home and so on.

Section 3 Consequences of decisions to move

22. With *hindsight*, do *you* think the decision that was made was the *right decision*?

    Prompts: Please explain.

23. What *differences* has moving here *made to your life*?

    Prompts: What improvements? What drawbacks? What has stayed the same?

24. Do you think you can *be yourself* here?

    Prompt: Please explain.

25. *What is it like* living here?

    Prompts: Typical daily life, social life, leisure and other activities, warden, care staff, other residents (tenants or owners?), amount and quality of care and support services, financial implications, making complaints, residents' committees, contact with local community and family/friends.

26. What would *improve your experience* of living here?

    Prompts: Improve specific housing related things.

27. What would *improve quality of life* in general?

    Prompts: anything else?

Section 4 About you

We would like to make sure that we speak to a range of older people. The details you give here will only be used for statistical purposes.

1. Name:
2. Are you: Please √: Female ☐ Male ☐
3. How old are you? *Please write your age in the boxes:* ☐☐
4. Are you:

   ☐ Single (never married)
   ☐ Married
   ☐ Widowed
   ☐ Living with a partner
   ☐ Divorced/separated

5. Do you consider yourself to be: *Please √ one box.*

   ☐ White
   ☐ Black-Caribbean
   ☐ Black-African
   ☐ Black-other
   ☐ Indian
   ☐ Pakistani
   ☐ Bangladeshi
   ☐ Chinese

   Any other ethnic group (*please describe*) ....

6. Before you retired, what was your

   occupation ....

7. Which of the following do you receive (or that you have) now? Please √ *all that apply*

   ☐ State pension ☐ Income support
   ☐ Private pension ☐ Disability benefit
   ☐ Occupational pension ☐ Housing benefit
   ☐ Savings/Investments ☐ Mobility benefit
   ☐ Family contributions ☐ Attendance Allowance
   ☐ Other benefit or income

   (please describe) ....

8. Do you have any limiting long-term illness(es) or disabilities?

☐ Yes
☐ No

If Yes, please describe . . . .

9. Do you feel that you need help in any of the following? If so, in what ways?

☐ Seeing . . . .
☐ Hearing . . . .
☐ Memory . . . .
☐ Mobility . . . .
☐ Washing and bathing . . . .
☐ Dressing and undressing . . . .
☐ Eating and drinking . . . .
☐ Using the toilet . . . .

10. Do any of the following usually live in your household with you?

Please √ all that apply

I Live Alone

Mother/ father

Wife/husband/partner

Daughter-in-law/son-in–law

Son

Daughter

Friend(s)

Sister/brother

Strangers/people I don't know well

Other . . . .

11. Do you have any close friends or relatives living nearby whom you can call on for help?

Yes ☐ No ☐

12. Please √ the box that best describes your accommodation:

☐ Owner-occupier(s) (outright or with a mortgage or loan)
☐ Renting (private landlord or Housing Association)
☐ Renting (from the Council)
☐ Sheltered housing without resident warden

- ☐ Sheltered housing with resident warden
- ☐ Very sheltered housing (e.g. warden + extra services–supported housing)
- ☐ Residential home
- ☐ Nursing home
- ☐ Other (*please describe*) ....

13. What type of dwelling do you live in?

    House   Bungalow   Room only   Flat   Other ....

14. How many bedrooms do you have?   1   2   3   4   5+

    (Please circle)

Appendix 2: Questionnaire

Housing decisions in old age: questionnaire

Instructions:

1) Please read and complete all the questions, unless instructed otherwise. Not all sections apply to everyone.

2) Some questions need you to tick all answers that apply. Others just need one answer. To help you, these questions have their own individual instructions.

3) Section 8 of this questionnaire is particularly important to help us make sense of the other information you have given. We would be most grateful if you could check that you have completed question 8 before you return the form.

4) When you have completed this questionnaire, please return it to **Liz Brooks, Counsel and Care, Twyman House, 16 Bonny Street, London, NW1 9PG** in the pre-paid envelope provided.

Section 1. Housing decision-making

1.1 *How long* have you lived in your *present home*? . . . . years *or* . . . . . months
1.2 Which statement below *most* applies to your own situation?
(PLEASE TICK ONLY ONE)

☐ I have *not even considered* moving
☐ I have thought about moving but have *decided to stay put*
☐ I have thought about moving but *feel daunted* by the *upheaval*
☐ I have thought about moving but *feel unsure* about the *pros* and *cons*
☐ I *would like to move* but do *not* feel I have *any suitable options*
☐ I *intend to move* but have *not decided where* to move to
☐ I am in the *process of moving*
☐ I have *recently moved*

1.3 How long *have you been* thinking about moving?
☐ Not at all ☐ Less than a month ☐ Less than a year
☐ A few years ☐ Many years

1.4 In making a housing decision, how much consideration did you, or would you, give to the following reasons?

(PLEASE ANSWER EACH QUESTION)

Reasons	**A lot**	**Some**	**Very little**	**None**
a) To be nearer particular *people* (e.g. friends, family)				
b) To be nearer particular *services* (e.g. hospital, leisure centre)				
c) To more *suitable housing* to meet your needs as you grow older				
d) To give you more *money* for other purposes				
e) To give you more *time* to pursue leisure activities (e.g. less housework)				
f) To be in an *area* you *know well* (e.g. through holidays, previously lived in)				
g) To move *back to your roots* (e.g. country of birth or area you were brought up in)				

1.5 *How important* were, or might the following be, in *influencing your decision whether or not to move?*

(PLEASE ANSWER EACH QUESTION)

Feelings	**Very important**	**Quite important**	**Neither**	**Quite unimportant**	**Very unimportant**
a) Feeling lonely					
b) Having an accident and nobody knowing					
c) Having an accident and nobody available to help					
d) Feeling unsafe in your house					
e) Feeling unsafe in your street/or your neighbourhood					
f) Feeling unable to cope with managing the house/garden					
g) Feeling unable to look after yourself					
h) Feeling isolated and unable to get out and about					

1.6 What kinds of *information* or *advice* would help you *to make decisions about your accommodation*?

Vital	Useful	Unnecessary	
☐	☐	☐	How to *adapt your house* to enable you to stay put
☐	☐	☐	*Grants/services* that are *available* to help you stay put
☐	☐	☐	*Questions to ask* when considering housing options
☐	☐	☐	*Differences* between various *housing options*
☐	☐	☐	*Housing options* within your *desired area*
☐	☐	☐	*Independent, tailor-made* housing advice

1.7 Which *type of organisation or person* would you prefer to go to for advice?

(PLEASE TICK if YES or NO)

Yes	No	
☐	☐	*Voluntary/charity* (e.g. Age Concern or Citizen's Advice Bureau)
☐	☐	*Home Improvement Agencies* (e.g. "Staying Put" or "Care & Repair")
☐	☐	Local *community* or *religious leader*
☐	☐	*Local council* worker (e.g. Housing officer)
☐	☐	*Housing association* or *private company* selling housing
☐	☐	*Estate agent*

1.8 How strongly do you *agree* or *disagree* with each of the following?
(PLEASE ANSWER EACH QUESTION)

Decision-making approach	**Strongly agree**	**Agree**	**Neither**	**Disagree**	**Strongly disagree**
a) "I wish to plan now for possible problems which may affect my ability to stay put"					
b) "I prefer to stay in my home until it becomes impossible to stay"					
c) "I would like to start thinking about my future housing needs, but I don't know where to start"					
d) "I would prefer to react to any housing problems as they arise"					
e) "There is no point planning for things that may never happen"					
f) "I do not want to think about my future"					
g) "I don't have the energy to move house at my age"					
h) "You can't plan for things that only might happen"					
i) "I don't know enough about what the options are to make an informed decision"					
j) "I would consider moving before things became difficult, to avoid having to move in a crisis"					

1.9 How *often* have you *moved house* in your life?

☐ Often (more than 10 times) ☐ Sometimes (6–10 times)
☐ Rarely (between 1 and 5 times) ☐ Never

Section 2. House moves in later life

If you have NOT moved since retirement or after the age of 60 years, please go to Section 3. If you have made more than one move since retirement, please let us know about your most recent move.

2.1a) *What was the* main reason(s) *for your last accommodation move?*

2.1b) If you have given more than one reason, can you say which was the *single, over-riding reason* for the move?

2.2 Do you consider your *last housing move* was?
(PLEASE SAY HOW MUCH YOU AGREE OR DISAGREE WITH EACH OF THESE STATEMENTS)

Statement	**Strongly agree**	**Agree**	**Neither**	**Disagree**	**Strongly disagree**
a) "A reaction to circumstances"					
b) "A decision taken with prevention in mind"					
c) "The result of a long period of planning"					
d) "Taken spontaneously"					
e) "Chosen willingly"					
f) "Chosen reluctantly"					
g) "The result of a crisis"					
h) "The result of longer term problems"					

2.3 Was *anyone else involved* in making the decision about where you should live?
(PLEASE TICK ALL THAT APPLY)

☐ Husband/wife/partner
☐ My child/children
☐ Social worker
☐ Doctor
☐ Other health professional

☐ Landlord
☐ Other

2.4 Who was the *most influential person* in making the decision?
(PLEASE TICK ONLY ONE)

☐ Self
☐ Husband/wife/partner
☐ Joint equal decision (i.e. between husband/wife/partner and myself)
☐ My child/children
☐ Social worker
☐ Doctor
☐ Other health professional
☐ Landlord
☐ Other

2.5 Was the decision t*o move out* of your previous accommodation the right one?

☐ Yes ☐ No ☐ Unsure

2.6 Was the decision *to move here* (to your present address) the right one (or would elsewhere have been better)?

☐ Yes ☐ No ☐ Unsure

2.7 In relation to *other major decisions* you have had to make in your life, how *stressful* is making a *housing decision in later life*?
(PLEASE CIRCLE THE SCORE)

Least stressful ► Most stressful

1 2 3 4 5

Section 3. Your present home

3.1 Which of the following *types of accommodation* best describes your present home?
(PLEASE TICK ONLY ONE)

☐ *Bungalow*
☐ *High-rise flat* or *maisonette* (your flat is on 5th floor or higher up)
☐ *Low-rise flat* (ground floor flat, or flat on 1st, 2nd, 3rd, or 4th floor)
☐ *House*
☐ Room in a *shared* house, room in a *flat-share*
☐ *Sheltered housing, without warden* (i.e. Bed-sitting room/flat/bungalow with personal alarm system)
☐ *Sheltered housing, with warden* (i.e. Bed-sitting room/flat/ bungalow)
☐ *Sheltered housing, with warden* and *extra support services* (i.e. Bed-sitting room/flat/bungalow)

☐ Room in a *residential care home*
☐ Room in a *nursing home*
☐ Other *(please specify)* ....

3.2 Which of the following *best describes* your present accommodation situation?
(PLEASE <u>TICK ONLY ONE</u>)

☐ Living with *family or friends* (don't own)
☐ *Own* outright (mortgage paid off in full)
☐ *Own* outright with *equity release scheme*
☐ *Own* with *mortgage*
☐ *Rent* from a housing association/housing charity
☐ *Rent* from council
☐ *Rent* from a private landlord/company
☐ Other *(please specify)* ....

3.3 How many *bedrooms* does your home have?

Bed-sitting room ☐
1 ☐
2 ☐
3 ☐
4 or over ☐

3.4 Do any of the following people usually *live in your home with you*?
(PLEASE TICK <u>ALL THAT APPLY</u>)

☐ I live alone
☐ Husband/wife/partner
☐ Mother/father
☐ Son
☐ Daughter
☐ Daughter-in-law/son-in-law
☐ Sister/brother
☐ Friend(s)
☐ People who are neither friends nor relatives (e.g. lodger, co-tenant)

Section 4. Suitability of present home

4.1 Given your present circumstances, would you describe your home as *suitable for you*?

☐ Yes ☐ No ☐ Unsure

4.2 How do you feel about the amount of *space* you have *in your home*?
(PLEASE <u>TICK ONLY ONE</u>)

☐ "I've got *too much* space, but I *do not regard this as a problem*"
☐ "I've got *too much* space. I would *prefer to have less*"
☐ "The space I have is *about right* for my everyday needs and hobbies"
☐ "I *don't have enough space* for my everyday needs and hobbies"

4.3 How many *steps or stairs* do you *have* to climb either to *reach* your own accommodation, *outside* or *within* your accommodation?

☐ One step
☐ Have lift and steps/stairs
☐ None
☐ Short flight of steps/stairs (up to 13)
☐ Long flight of steps/stairs (14 or more)

4.4 Are *stairs* a problem for you *within* your home?
(PLEASE TICK ONLY ONE)

☐ Yes, I have difficulty using the stairs, so I use the upstairs of my house less
☐ Yes, I have difficulty using the stairs, but I manage
☐ I have a stair lift, so there isn't a problem
☐ No problems with stairs
☐ I have no stairs within my home

4.5 How satisfied are you with the *condition of your present home*?

Housing condition	**Completely satisfied**	**Somewhat satisfied**	**Neither**	**Somewhat dissatisfied**	**Completely dissatisfied**
Internal condition (e.g. damp, condensation)					
External condition (e.g. major repairs)					

4.6 Do you have *any problems* with doing any of the following jobs *yourself* or *getting others to do them* for you?
(PLEASE ANSWER EACH QUESTION)

Household	**A lot (e.g. I can't manage alone *and* I can't get/ afford the help I need)**	**Some (e.g. I can't manage alone, but I can get/ afford the help I need)**	**A few (e.g. I find these jobs difficult and/or expensive so they don't get done)**	**No problems**
External housing maintenance (e.g. roof repairs, window replacement)				
Internal housing maintenance (e.g. decoration, D-I-Y)				
Odd jobs (e.g. changing light bulbs, plugs)				
Garden maintenance				
Heavy housework (e.g. hovering, spring cleaning)				
Administration, (e.g. paying bills, getting people to carry out servicing and repairs)				

4.7 Apart from what you do yourself, how much help do you get from other people to *help you look after your home*?
(PLEASE ANSWER EACH QUESTION)

Help from others	**Frequent (at least once a day)**	**Regular (at least once a week)**	**Occasional (at least once a month)**	**Emergency**
Husband/wife/partner				
Family				
Friends/neighbours				
Volunteers				
Tradespeople				
Home help				

4.8 Which of the following do you have which you could use to call for *emergency help*?
(PLEASE TICK ALL THAT APPLY)

☐ mobile phone/cordless phone ☐ telephone ☐ personal alarm
☐ alarm button/or cord in room(s) ☐ none of these

Section 5. Suitability of present neighbourhood

5.1 *How often* do you use these *forms of transport* to go *shopping* or generally *getting out* and *about*?

(PLEASE ANSWER EACH QUESTION)

Transport	Often	Sometimes	Rarely	Never
Walking				
Walking with mobility aid				
Wheelchair				
Buggy or motorised wheelchair				
Cycling				
Own car – drive myself				
Own car – have to be driven				
Car lifts from family/friends				
Dial-a-ride scheme				
Voluntary transport scheme or similar				
Buses				
Taxis				
Trains				

5.2 Do you have *any problems getting to* or *from* any of the following places? **(PLEASE ANSWER EACH QUESTION)**

Places	**A lot**	**Some**	**Very little**	**None**	**Does not apply**
Bus stop					
Medical services (e.g. GP, health centre)					
Hospital					
Essential shops (e.g. chemist, food shops)					
Banking services (e.g. bank, post office)					
Close family and friends					
Leisure and entertainment services (e.g. classes, cinema, clubs, library)					
Other shops					
Place of religious worship					

5.3 How regularly do you use *local public transport*?

☐ At least once a day
☐ At least once a week
☐ At least once a month
☐ Once a year or less
☐ I do not use public transport

5.4 How satisfied are you that your *local public transport gets you to where you want to go*?

☐ Completely satisfied
☐ Mostly dissatisfied
☐ Mostly satisfied
☐ Completely dissatisfied
☐ Neither satisfied nor dissatisfied
☐ Public transport does not meet my needs, so I do not use it at all

5.5 How *safe* do you feel *in your own home*?

☐ Completely safe ☐ Quite safe ☐ Quite unsafe ☐ Not safe at all

5.6 How *safe* do you feel when you are *outside your home*?
(i.e. in your *local neighbourhood*?)

☐ Completely safe ☐ Quite safe ☐ Quite unsafe ☐ Not safe at all

Section 6. Future housing decision-making

6.1 How much *time have you spent thinking* about *whether your home or neighbourhood will still be suitable,* should your circumstances change?

☐ A lot ☐ Some ☐ Very little ☐ None

6.2 *If* any of the following happened, would they affect your *ability or wish* to stay where you are?
(PLEASE ANSWER EACH QUESTION)

Change in circumstances	**Definitely a problem**	**Possibly a problem**	**Unlikely to be a problem**	**Would be a problem**
a) Your husband/ wife/ partner becomes *seriously ill*				
b) Your husband/ wife/ partner *died*				
b) You needed to use a *wheelchair* or other *mobility aids*				
c) You could no longer manage the *stairs*				
d) You were less able to *afford* the bills or repairs				
e) You were less able to look after the *house/garden*				
f) People who presently offer you *support* were less able to help				
g) Local *facilities* closed				
h) You were no longer able *to drive*				
i) Your local *transport* systems got worse				
j) Your *neighbourhood* got worse, e.g. too noisy, did not feel safe				

6.3 How *confident* are you that you will be able to *have sufficient control* over *any future decision* about moving?

☐ Very confident
☐ Quite confident
☐ Not at all confident
☐ Unsure

6.4 Which of these *housing options* would you consider *now*, if they were available in your local neighbourhood? *(By neighbourhood, we mean up to approximately 15-minute walk from your present home.)*
(PLEASE ANSWER EACH QUESTION)

Housing options you would consider NOW	**Definitely**	**May be**	**Reluctantly**	**Would not consider**	**Already living in this accomo-dation**
a) Flat/bungalow with alarm system but without warden					
b) Flat/bungalow with warden					
c) Flat/bungalow with warden & extra support services					
d) Residential home					
e) Nursing home					
f) Staying put with adaptations and support services					

6.5 Imagine in the future:

- you were *living alone* (even if now you live with a husband/wife/partner) AND
- you had a *serious disability* or *serious long-term illness* that severely limits your ability to look after yourself and your home

In such a situation, which of these options would you consider if they were available in your local neighbourhood?

(PLEASE ANSWER EACH QUESTION)

Housing options you would consider IN THE FUTURE	**Definitely**	**May be**	**Reluctantly**	**Would *not* consider**	**Already living in this accommo-dation**
a) Flat/bungalow with alarm system but without warden					
b) Flat/bungalow with warden					
c) Flat/bungalow with warden and extra support services					
d) Residential home					
e) Nursing home					
f) Staying put with adaptations and support services					

6.6a What would *put you off moving* into housing that is *specifically designed* for *older people's needs?*

(PLEASE ANSWER EACH QUESTION)

What would put you off?	**Strongly agree**	**Agree**	**Neither**	**Disagree**	**Strongly disagree**
a) "If most of the current residents were *a lot older than me*"					
b) "If I had to *move out of my neighbourhood*"					
c) "If I had to *move away from the geographical area* I consider home"					
d) "The *private living space* within older people's housing tends to be *too small*"					
e) "I don't like the idea of *community living*"					
f) "I don't like the idea of living *only with other older people*"					
g) "There might be too many *rules and regulations*"					
h) "Schemes might *not be culturally sensitive* (e.g. staffing, food)"					
i) "I might be living alongside *people I would not normally choose to live with* (e.g. they don't understand my culture, religion)"					

6.6b *Is there anything else which might put you off housing that is specifically designed for older people's needs?*

(Please specify) ....

Section 7. Your health and well-being

7.1 In general, in comparison with other people of your age, would you say your *general state of health* is?

☐ Excellent
☐ Very good
☐ Fair
☐ Not very good
☐ Poor

7.2 Do you have any *long-standing health problems* that you are experiencing that *adversely affect or restrict your way of life* at the moment?
(By long standing, we mean anything that has troubled you over a period of time or is likely to affect you over a period of time – e.g. arthritis, heart trouble, breathing problems).

☐ Yes (Please specify) ....
☐ No

7.3 *Do you need or receive regular help with the following?*

Help from others	**Frequent (at least once a day)**	**Regular (at least once a week)**	**Occasional (at least once a month)**	**Emergency**
Personal care (e.g. bathing, dressing,getting in and out of bed)				
Nursing care (e.g. changing dressings,giving injections)				

7.4 How do you feel about your **present quality of life?**

☐ Completely satisfied
☐ Mostly satisfied
☐ Neither satisfied or dissatisfied
☐ Mostly dissatisfied
☐ Completely dissatisfied

7.5 How important is your *housing* in relation to your *overall quality of life?*

☐ Very important
☐ Quite important
☐ Neither important nor unimportant
☐ Quite unimportant
☐ Very unimportant

Section 8. Some questions about you

8.1 Are you? ☐ Male ☐ Female

8.2 Are you?

☐ Single (never married)
☐ Married or living with partner
☐ Separated
☐ Divorced
☐ Widowed or widower

8.3 How old are you? .... Years

8.4 Which of the following best describes your situation?

☐ I was in paid employment and am *now retired* *Please go to next question*
☐ I was not in paid employment (e.g. caring for dependants etc.) *Please go to 8.6*
☐ I have not retired yet, I work *full time* *Please go to 8.6*
☐ I have not retired yet, I work *part time* *Please go to 8.6*

8.5 What *age did you retire*? .... Years

8.6 What was or is your *occupation* (or occupation of head of household?)
....

8.7 What is your *ethnic group*?
Choose one section from A to E, then tick the appropriate box to indicate your background.

A. White British

☐ White English ☐ White Scottish
☐ White Welsh
☐ White Irish ☐ White other, please write in ....

B. Mixed

☐ White and Black-Caribbean
☐ White and Black-African, White and Asian
☐ Any other Mixed background, please write in ....

C. Asian, Asian British, Asian English, Asian Scottish, or Asian Welsh

☐ Indian ☐ Pakistani
☐ Bangladeshi
☐ Any other Asian background, please write in ....

D. Black, Black British, Black English, Black Scottish, or Black Welsh

☐ Caribbean ☐ African
☐ Any other Black background, please write in ....

E. Chinese, Chinese British, Chinese English, Chinese Scottish, or Chinese Welsh or other ethnic group

☐ Chinese ☐ Any other background, please write in ....

8.8 Are you in receipt of any of the following?
(PLEASE TICK ALL THAT APPLY)

☐ Earnings from employment/self-employment
☐ Company Pension
☐ Private Pension
☐ Interest from savings/investments
☐ State Pension
☐ Housing Benefit
☐ Attendance Allowance
☐ Disability Living Allowance
☐ Income from another source

8.9 What group represents your (and your husband/wife's) total net income from all these sources after deductions for income tax and National Insurance? If unsure, please estimate
(PLEASE TICK ONLY ONE)

Weekly	*or*	*Calendar-monthly*
☐ Under £60		☐ Less than £260
☐ £60–£99		☐ £260–£429
☐ £100–£159		☐ £430–£689
☐ £160–£199		☐ £690–£869
☐ £200–£299		☐ £870–£1299
☐ £300–£399		☐ £1300–£1729
☐ £400+		☐ £1730+

Section 9. And finally

9.1 Did anyone else help you fill in this questionnaire and if so, who?

☐ No, I filled it in myself
☐ Son
☐ Daughter
☐ Sister/brother
☐ Wife/husband/partner
☐ Daughter-in-law/son-in-law
☐ Friend(s)
☐ Neighbour(s)
☐ Other ....

9.2 Thinking about your experience of making housing decisions, do you have any suggestions or ideas on what would make this decision-making easier?

Thank you for taking the time to fill in this questionnaire. It is very much appreciated.
A reply by *May 15 2002* entitles you to entry in our *Prize Draw*.
One top prize of £300;
Two second prizes of £100;
Three third prizes of £50.

Appendix 3: Suggested 'Letter to the Editor' to Ask for Written Housing Stories

Your chance to play a part in improving housing for older people

Here's your chance to play your part in an important research study, which aims to help improve the quality of life of people in retirement. Lancaster University and Counsel and Care for the Elderly are trying to find out how older people make decisions about housing, and how those decisions affected their well-being.

We would like people who are facing making decisions about where to live in retirement now, or have already made this decision, to write and tell us about it (or use a tape recorder if that is easier). This might have meant making changes to your existing home, or having to move house.

The kind of questions we would like answered are:

- What led you into thinking about the kind of accommodation you need as you get older?
- The reasons you have for staying put, or moving house?
- What you are especially looking for in your housing during retirement?
- If you cannot get everything you would like, what would you consider your essential housing needs?
- The kind of help you would welcome (or did you get) in making your decision?
- If you have already made your decision, with hindsight, tell us whether you feel it was the right one, and why or why not.

These are some suggestions, but please feel free to tell us anything else you wish about your housing decisions. What you write will help provide better housing for older people in the future.

Les Bright
Deputy Chief Executive
Counsel and Care

Appendix 4: Written Housing Stories Press Release

Trade press

Community Care
This Caring Business
Elderly Care
Hospital Doctor
Therapy Weekly
Nursing and Residential Home News
Nursing Times & Mirror
Nursing Standard

Local releases

London

sent to all local weekly newspapers in Greater London and

Evening Standard
GLR
London Live (radio)
Capital
Carlton/LWT
BBC Television South East

Lancaster

sent to all weekly papers in Lancaster and

Lancashire Evening Post
Lancashire Evening Gazette
Evening Telegraph, Blackburn
Manchester Evening News
Evening Post Wigan
Evening Chronicle, Oldham
Bolton Evening News
Red Rose Radio
BBC Radio Lancashire
The Bay Radio
Radio Wave
Border TV
BBC Television North-West
Granada TV
BBC GMR

Housing associations, older people's charities and organizations

Abbeyfield
Age Concern
Anchor Housing
Care and Repair
Housing 21
Methodist Homes for the Aged
Retirement Security Ltd
Hanover Housing
Help the Aged
University of the Third Age
Over sixties clubs
Rural/parish newsletters – in selected locations

Notes

Chapter 2: Telling Older People's Housing Stories

1. Health, social and residential care (GP, nursing, social work, housekeeping, catering, home help); voluntary sector (prison visiting, victim support, citizen advice bureau, community health council, age concern, magistrate); industry and business (managerial positions, engineering, journalism); and the public sector (librarian, inland revenue, police force).
2. 'It's good to have someone to talk to', Finch, 1984.
3. 'Surveying through stories', Graham, 1984.

Chapter 3: Housing Decisions in Later Life

1. Unpublished comment in a seminar.

Chapter 9: Theory, Policy and Practice

1. Personal communication.

References

Afshar H., Franks M. and Maynard M. (2002) Women, ethnicity and empowerment in later life, *Go Findings 10*, Sheffield: Growing Older Programme.

Appleton N. (2002) *Planning for the Majority: The Needs and Aspirations of Older People in General Housing*, York: Joseph Rowntree Foundation.

Arber S. and Evandrou M. (eds) (1993) *Ageing, Independence and the Life Course*, London: Jessica Kingsley.

Arnstein S. (1969) Ladder of citizen participation, *Journal of the American Institute of Planners*, 35(4): 216–24.

Askham J., Nelson H., Tinker A. and Hancock R. (1999) *To Have and to Hold: the Bond Between Older People and the Homes They Own*, York: Joseph Rowntree Foundation.

Atkinson T. and Claxton G. (eds) (2001) *The Intuitive Practitioner: On the Value of Not Always Knowing What One Is Doing*, Buckingham: Open University.

Audit Commission (2004) *Assistive Technology: Independence and Well-being 4*, London: Audit Commission.

Baron J. (1994) *Thinking and Decision Making* (2nd edn), New York: Cambridge University Press.

Baudrillard J. (1988) *Jean Baudrillard: Selected Writings*, Cambridge: Polity Press.

Bauman Z. (1987) *Legislators and Interpreters*, Cambridge: Polity Press.

Bazerman M.H. (1994) *Judgements in Managerial Decision Making* (4th edn), New York: Wiley.

Beaumont G. and Kenealy P. (2003) Quality of life of healthy older people: residential setting and social comparison processes, *GO Findings 23*, Sheffield: Growing Older Programme.

Beck U. (1992) *Risk Society: Towards a New Modernity*, translated by Ritter M., London: Sage.

Berrryhill (2004) www.berryhillvillage.freeserve.co.uk.

Better Government for Older People (BGOP) (2002) *Annual Report 2001/02*, London: BGOP.

Biggs S. (1993) *Understanding Ageing*, Buckingham: Open University Press

Blakemore K. (1993) Ageing and ethnicity, in Johnson J. and Slater R. (eds), *Ageing and Late Life*, London: Sage.

Bland R. (1999) Independence, privacy and risk: two contrasting approaches to residential care for older people, *Ageing and Society* 19: 539–60.

Bocock R. (1993) *Consumption*, London: Routledge.

Bond J., Briggs R. and Coleman P. (1993) The study of ageing, in Bond J., Coleman P. and Peace S. (eds), *Ageing in Society*, London: Sage.

Bourdieu P. (1977) *Outline of a Theory of Practice*, translated by Nice R., Cambridge: Cambridge University Press.

Bourdieu P. (1984) *Distinction*, translated by Nice R., London: Routledge and Kegan Paul.

Bowling A., Grundy E. and Farquhar M. (1998) *Living Well into Old Age*, London: Age Concern.

Bowling A., Gabriel Z., Banister D. and Sutton S. (2003) Adding quality to quantity: older people's views on their quality of life and its enhancement, *Go Findings 7*, Sheffield: Growing Older Programme.
Brenton M. (1998) *'We're in Charge'. CoHousing communities of Older People in the Nederlands: Lessons for Britain*, Bristol: Policy Press.
Brenton M. (2002) Choosing and managing your own community in later life in Sumner K. (ed.) *Our Homes, Our Lives: Choice in Later Life Living Arrangements*, London: Centre for Policy on Ageing.
Brown L. and Moore E. (1970) The urban migration process: a perspective, *Geografiska Annaler*, 52B: 1–13.
Butt J., Moriarty J., Brockmann M., Chih Hoong Sin and Fisher M. (2003) Quality of life and social support among older people from different ethnic groups, *GO Findings 20*, Sheffield: Growing Older Programme.
Callaway H. (1992) Ethnography and experience: gender implications in fieldwork and texts, in Okely J. and Callaway H. (eds), *Anthropology and Autobiography*, London: Routledge.
Casciani, D. (2003) 'Fear of crime trapping elderly' Story from BBC NEWS http://news.bbc.co.uk/go/pr/fr/-/1/hi/uk/3044625.stm Published: 2003/05/20 23:13:05 GMT.
Chaney (1997) *Lifestyles*, London: Routledge.
Church K. (1995) *Forbidden Narratives: Critical Autobiography as Social Science*, Canada: Gordon and Breach Publishers.
Clapham D., Means R. and Munro M. (1993) Housing, the life course and older people, in Arber S. and Evandrou M. *Ageing, Independence and the Life Course*, London: Jessica Kingsley.
Claxton G. (2001) The anatomy of intuition, in Atkinson T. and Claxton G. (eds), *The Intuitive Practitioner: On the Value of Not Always Knowing What One is Doing*, Buckingham: Open University.
Clough (1981) *Old Age Homes*, London: Allen and Unwin
Clough R. (1993) Housing and services for older people, in Day P. (ed.) (1993) *Perspectives on Later Life*, London: Whiting and Birch.
Clough R. (1998) *Living in Someone Else's Home*, London: Counsel and Care.
Clough R. (2000) *The Practice of Residential Work*, Basingstoke: Macmillan.
Clough R. (2002) Commentary 2, in Sumner K. (ed.), *Our Homes, Our Lives: Choice in Later Life Living Arrangements*, London: Centre for Policy on Ageing.
Clough R., Leamy M. and Bright L. with Miller V. and Brooks L. (2003) *Homing in on Housing: a Study of Housing Decisions of People Aged over 60*, Lancaster: Eskrigge Social Research.
Clough R., Bright L., Green B., Hawkes B. and Raymond G. (2004) *Older People as Researchers: Potential, Practicalities and Pitfalls*, York: Joseph Rowntree Foundation.
Cobb S. (1976) Social support as a moderator of life stress, *Psychosomatic Medicine*, 38: 300–14.
Cooley C. (1922) *Human Nature and the Social Order*, revised edition, New York: Scribner's Press.
Croucher K., Pleace N. and Bevan B. (2003) *Living at Hartrigg Oaks: Residents' Views of the UK's First Continuing Care Retirement Community*, York: Joseph Rowntree Foundation.
Cumming E. and Henry W. (1961) *Growing Old: the Process of Disengagement*, New York: Basic Books.

Dalley G. (2002) Independence and autonomy: the twin peaks of ideology, in Sumner K. (ed.), *Our Homes, Our Lives: Choice in Later Life Living Arrangements*, London: Centre for Policy on Ageing.

DaVanzo J. (1981a) Repeat migration, information costs and location-specific capital, *Population and Environment*, 4(1): 45–73.

DaVanzo J. (1981b) Microeconomic approaches to studying migration decisions, in DeJong G. and Gardner R. (eds), *Migration Decision Making: Multidisciplinary Approaches to Microlevel Studies in Developed and Developing Countries*, New York: Pergamon Press.

de Certeau M. (1984) *The Practice of Everyday Life*, Berkeley, CA: University of California Press.

Dewar B., Jones C. and O'May F. (2004) *Involving Older People: Lessons for Community Planning*, Edinburgh: Scottish Executive.

De Koning K. and Martin M. (1996) Participatory research in health: setting the context in De Koning K. and Martin M. (eds), *Participatory Research in Health: Issues and Experiences*, London: Zed Books.

Department of Health (2001) *National Service Framework for Older People*, London: Department of Health.

Despres C. (1991) The meaning of home: literature review and directions for further research and theoretical development, *Journal of Architectural and Planning Research*, 8(2): 96–115.

Dittmar H. (1992) *The Social Psychology of Material Possessions*, New York: St. Martin's Press.

Douzinas C. (2002) Identity, recognition, rights or what can Hegel teach us about human rights? *Journal of Law and Society*, 29(3): 379–405.

Dupuis A. and Thorns D. (1996) The meaning of home for older homeowners, *Housing Studies*, 11(4): 485–501.

Easterbrook L., Horton K., Arber S. and Davidson K. (2001) *International Review of Interventions in Falls among Older People*, London: Department of Trade and Industry.

Einagel V. (2002) Telling stories, making selves, in Truman C., Mertens D. and Humphries B. (eds), *Research and Inequality*, London: UCL Press.

Erikson E. (1950) *Childhood and Society*, New York: Norton.

Evan R. (1981) The relationship of two measures of perceived control to depression, *Journal of Personality Assessment*, 45: 66–70.

Evans J. and Over D. (1997) Are people rational? Yes, no and sometimes, *The Psychologist*, 10(9): 403–6.

Festinger L. (1954) A theory of social comparison processes, *Human Relations*, 7: 117–40.

Finch J. (1984) 'It's good to have someone to talk to', in Bell C. and Roberts H. (eds), *Social Researching: Politics, Problems, Practice*, London: Routledge.

Fokkema T. and Van Wissen L. (1997) Moving plans of the elderly: a test of the stress-threshold model, *Environment and Planning*, 29(3): 249–68.

Folkman S., Lazarus R., Dunkel-Schetter C. and Delongis A. (1986) Appraisal, coping, health status and psychological symptoms, *Journal of Personality and Social Psychology*, 50: 571–9.

Forrest R. and Kemeny J. (1984) *Careers and Coping Strategies: Micro and Macro Aspects of the Trend Towards Owner-Occupation*, Mimeo, Bristol: University of Bristol.

Forrest R. and Murie A. (1985) *The Housing Histories of Home Owners: Some Preliminary Observations*, Mimeo, Bristol: University of Bristol.

Forrest R. and Murie A. (1991) Housing markets, labour markets and housing histories, in Allan J. and Hamnett C. (eds), *Housing and Labour Markets*, London: Unwin Hyman.

Furedi F. (2002) *The Culture of Fear: Risk-Taking and the Morality of Low Expectation*, London: Continuum.

Gandhi K. (2002) Commentary, in Sumner K. (ed.), *Our Homes, Our Lives: Choice in Later Life Living Arrangements*, London: Centre for Policy on Ageing.

Gardner H., Kornhaber M. and Wake W. (1996) *Intelligence: Multiple Perspectives*, Fort Worth: Harcourt Brace.

Gergen K. (1994) *Realities and Relationships: Soundings in Social Construction*, London: Harvard University Press.

Giddens A. (1990) *The Consequences of Modernity*, Cambridge: Polity Press.

Giddens A. (1991) *Modernity and Self-Identity: Self and Society in the Late Modern Age*, Cambridge: Polity Press.

Gigerenzer G., Tood P. and the ABC research group (eds) (1999) *Simple Heuristics that Make Us Smart*, New York: Oxford University Press.

Gilleard C. (1996) Consumption and identity in later life: toward a cultural gerontology, *Ageing and Society*, 16: 489–98.

Giorgi A. and Giorgi B. (2003) Phenomenology, in Smith J. (ed.), *Qualitative Psychology: a Practical Guide to Research Methods*, London: Sage.

Goodman J. (1981) Information, uncertainty and the microeconomic model of migration decision making, in DeJong G. and Gardner R. (eds), *Migration Decision Making: Multidisciplinary Approaches to Microlevel Studies in Developed and Developing Countries*, New York: Pergamon Press.

Graham H. (1984) Surveying through stories, in Bell C. and Roberts H. (eds), *Social Researching: Politics, Problems, Practice*, London: Routledge.

Gubrium J. (1993) *Speaking of Life: Horizons of Meaning for Nursing Home Residents*, New York: Aldine de Gruyter.

Gubrium J. and Lynott R. (1983) Rethinking life satisfaction, *Human Organisation*, 42: 30–8.

Gurney C. and Means R. (1993) The meaning of home in later life, in Arber S. and Evandrou M. (eds), *Ageing, Independence and the Life Course*, London: Jessica Kingsley.

Havighurst R. (1963) Successful ageing, in Williams R., Tibbitts C. and Donahue W. (eds), *Processes of Ageing*, vol. 1, New York: Atherton Press.

Hegel G. (1977) *Phenomenology of Spirit*, translated by Miller A., Oxford: Clarendon Press.

Hepworth M. (2000) *Stories of Ageing*, Buckingham: Open University Press.

Hermans H. (1999) Self-narrative as meaning construction, *Journal of Clinical Psychology*, 55(10): 1193–211.

Heywood F., Oldman C. and Means R. (2002) *Housing and Later Life*, Buckingham: Open University Press.

Heywood F., Pate A., Means R. and Galvin J. (1999) *Housing Options for Older People: Report on a Developmental Project to Refine a Housing Option Appraisal Tool For Use by Older People*, Bristol: School for Policy Studies.

Hormuth S. (1990) *The Ecology Of The Self*, Cambridge: Cambridge University Press.

Jefferson T. and Holloway W. (2000) The role of anxiety in the fear of crime, in Hope T. and Sparks R. (eds), *Crime, Risk and Insecurity*, London: Routledge.

Jerrome D. (1992) *Good Company: an Anthropological Study of Old People in Groups*, Edinburgh: Edinburgh University Press.

Johnson-Carroll K., Brandt J. and McFadden J. (1995) Factors that influence pre-retirees' propensity to move at retirement, *Journal of Housing for the Elderly*, 11(2): 85–105.

Kahneman D. and Tversky A. (1979) Prospect theory: an analysis of decision under risk, *Econometrica*, 47: 263–91.

Kahneman D., Slovic P. and Tversky A. (1982) *Judgement under Uncertainty: Heuristics and Biases*, Cambridge: Cambridge University Press.

Kellaher L. (2000) *A Choice Well Made: 'Mutuality' as a Governing Principle in Residential Care*, London: Centre for Policy on Ageing.

Kellaher L. (2002) Is genuine Choice a Reality in Sumner K. (ed.) (2002) *Our Homes, Our Lives: Choice in Later Life Living Arrangements*, London: Centre for Policy on Ageing.

Klein G. (1998) *Sources of Power: How People Make Decisions*, Cambridge, MA.: MIT Press.

Kohn M. (1972) Class, family and schizophrenia, *Social Forces*, 50: 295–302.

Lach W. (2002/03) Fear of falling: an emerging public health problem, *Generations – Journal of the American Society on Aging*, 26(4): 33–7.

LaPiere R. (1934) Attitudes and Actions, *Social Forces*, 13: 230–7.

Lash S. (1994) Reflexivity and its doubles: structure, aesthetics, community, in Beck U., Giddens A. and Lash S., *Reflexive Modernization*, Cambridge: Polity Press.

Lash S. and Urry J. (1987) *The End of Organised Capitalism*, Cambridge: Polity Press.

Laws G. (1995) Embodiment and emplacement – identities, representation and landscape in Sun City retirement communities, *International Journal of Aging and Human Development*, 40(4): 253–80.

Lazarus R. and Folkman S. (1984) *Stress, Appraisal and Coping*, New York: Springer.

Leamy M. (2005) Helping older people to share the research journey, in Lowes L. and Hulatt I. (eds), *Service Users' Involvement in Health and Social Care Research*, London: Routledge

Leamy M. and Clough R. (2004) *Depending on Each Other: Teaching, Researching and Older People's Perspectives on Doing Research Together*, York: Joseph Rowntree Foundation.

Leder D. (1990) *The Absent Body*, London: University of Chicago Press.

Maslow A. (1987) *Toward a Psychology of Being*, London: Longman.

Mc Craith M. (2002) Creating a home, unpublished creative writing exercise.

McHugh K. (2000) The 'ageless self'? Emplacement of identities in sun belt retirement communities, *Journal of Aging Studies*, 14(1): 103–15.

Mc Intyre A. (1999) Elderly fallers: a baseline audit of admissions to a day hospital for elderly people, *British Journal of Occupational Therapy*, 62(6): 244–8.

Mead G. ([1934a] 1977a) *Mind, Self and Society: From the Standpoint of a Social Behaviourist*, ed. Morris C., Chicago: University of Chicago Press.

Mead G. ([1934b] 1977b) *On Social Psychology*, ed. Strauss A., Chicago: University of Chicago Press.

Means R. (1999) Housing and housing organisations: a review of their contribution to alternative models of care for elderly people. Appendix 3 of *Alternative Models of Care for Older People*, Research Volume 2 of the Royal Commission Report, London: Stationery Office.

Mirowsky J. and Ross C. (1984) Mexican culture and its emotional contradictions, *Journal of Health and Social Behaviour*, 25: 2–13.

Moody H. (1988) Towards a critical gerontology: the contribution of the humanities to theories of ageing, in Birren J. and Bengston V. (eds), *Emergent Theories of Ageing*, New York: Springer.

Murphy S., Dubin J. and Gill T. (2003) The development of fear of falling among community-living older women: predisposing factors and subsequent fall events, *Journals of Gerontology Series A-Biological Sciences and Medical Sciences*, 58(10): 943–7.

Nazroo J., Bajekal M., Blane D., Grewal I. and Lewis J. (2003) Ethnic inequalities in quality of life at older ages: subjective and objective components, *GO Findings 11*, Sheffield: Growing Older Programme.

Okely J. (1996) *Own or Other Culture*, London: Routledge.

Oldman C. (1991) Financial effects of moving in old age, *Housing Studies*, 6(4): 251–62.

Oldman C. and Quilgars D. (1999) The last resort? Revisiting ideas about older people's living arrangements, *Ageing and Society*, 19(4): 363–84.

Park D. and Schwarz N. (2000) *Cognitive Aging: A Primer*, Hove: Psychology Press.

Parker R. (1988) An historical background, in Sinclair I. (ed.) (1988) *Residential Care: The Research Reviewed*, London: HMSO.

Pastalan L. (ed.) (1995) Housing decisions for the elderly: to move or not to move, *Journal of Housing for the Elderly*, special edition, 11(2): 1–5.

Pastalan L. and Barnes J. (1999) Personal rituals: identity, attachment to place, and community solidarity, in Schwarz B. and Brend R. (eds), *Aging, Autonomy, and Architecture: Advances in Assisted Living*, Baltimore, MD: Johns Hopkins University Press.

Payne J. and Payne G. (1977) Housing pathways and stratification: a study of life chances in the housing market, *Journal of Social Policy*, 6(2): 129–56.

Peace S. (1999) *Involving Older People in Research: an Amateur Doing the Work of a Professional?*, London: Centre for Policy on Ageing.

Peace S. (2002) The role of older people in social research, in Jamieson A. and Victor C. (eds), *Researching Ageing and Later Life*, Buckingham: Open University.

Percival J. (2002) Domestic spaces: uses and meanings in the daily lives of older people, *Ageing and Society*, 22(6): 729–49.

Phillipson C. and Thompson N. (1997) The social construction of old age, in Bland R. (ed.) (1997) *Developing Services for Older People and their Families*, London: Jessica Kingsley.

Reason J. (ed.) (1988) *Human Inquiry in Action: Developments in New Paradigm Research*, London: Sage.

Reason P. (ed.) (1994) *Participation in Human Inquiry*, London: Sage.

Rochberg-Halton E. (1984) Object relations, role models, and cultivation of the self, *Environment and Behaviour*, 16(3): 335–68.

Rodin J. (1986) Ageing and health: effects of the sense of control, *Science*, 233: 1271–6.

Rogers W. (2003) *Social Psychology: Experimental and Critical Approaches*, Buckingham: Open University Press.

Rolfe S., Mackintosh S. and Leather P. (1993) *Age File '93*, Oxford: Anchor Housing Trust.

Rose N. (1999) *Governing the Soul: the Shaping of the Private Self*, London: Free Association Books.

Rotter J. (1966) Generalised expectancies for internal versus external control of reinforcement, *Psychological Monographs*, 80: 1–28.
Sapsford R. (2003) Domains of analysis, in Sapsford R., Still A., Wetherell M., Miell D. and Stevens R. (eds), *Theory and Social Psychology*, London: Sage.
Sarbin T. (1986) The narrative as a root metaphor for psychology, in Sarbin R. (ed.), *Narrative Psychology: the Storied Nature of Human Conduct*, New York: Praeger.
Schneider S. (1992) Framing and conflict: aspiration level contingency, the status quo and current theories of risky choice, *Journal of Experimental Psychology: Learning Memory and Cognition*, 18: 1040–57.
Schroots J. and Birren J. (1990) The nature of time, *Comprehensive Gerontology*, 2: 1–30.
Secker J., Hill R., Villeneau L. and Parkman S. (2003) Promoting independence: but promoting what and how? *Ageing and Society* 23: 375–91.
Seligman M. (1975) *Helplessness*, San Francisco, CA: Freeman.
Shafir E., Simonson I. and Tversky A. (1993) Reason-based choice, *Cognition*, 49: 11–36.
Silver I. (1996) Role transitions, objects, and identity, *Symbolic Interaction*, 19(1): 1–20.
Simon H. (1956) Rational choice and the structure of environments, *Psychological Review*, 63: 129–38.
Simon H. (1957) *Models of Man: Social and Rational*, New York: Wiley.
Sixsmith A. (1986) Independence and home in later life, in Phillipson C., Bernard M. and Strang P. (eds), *Dependency and Interdependency in Old Age: Theoretical Perspectives and Policy Alternatives*, London: Croom Helm.
Smyer M. and Qualls S. (1999) *Aging and Mental Health*, Oxford: Blackwell.
Speare A. (1971) A cost–benefit model of rural to urban migration in Taiwan, *Population Studies*, 25: 117–30.
Strauss A. and Corbin J. (1998) *Basics of Qualitative Research: Techniques and Procedures for Developing Grounded Theory*, London: Sage Publications.
Stroebe W. (2000) *Social Psychology and Health*, Buckingham: Open University Press.
Tasker L. (1978) Class, culture and community work, in Curno P. (ed.), *Political Issues and Community Work*, London: Routledge and Kegan Paul.
Taylor M., Hoyles L., Lart R. and Means R. (1992) *User Empowerment in Community Care*, London: Chapman and Hall.
Thagard P. (2001) How to make decisions: coherence, emotion, and practical inference, in Millgram E. (ed.), *Varieties of Practical Inference*, Cambridge, MA: MIT Press.
Tinetti M. and Powell L. (1993) Fear of falling and low self-efficacy: a cause of dependence in elderly persons', *The Journal of Gerontology*, 48: 35–8.
Tinker A. (1997) *Older People in Modern Society*, London: Longmans.
Torrington J. (2002) Commentary in Sumner K. (ed.), *Our Homes, Our Lives: Choice in Later Life Living Arrangements*, London: Centre for Policy on Ageing.
Tversky A. (1972) Elimination by aspects: a theory of choice, *Psychological Review*, 79: 281–99.
Urry, J. (2000) *Sociology Beyond Societies: Mobilities for the Twenty-First Century*, London: Routledge.

Vanderhardt P. (1995) The housing decisions of older households: a dynamic analysis, *Journal of Housing Economics*, 7: 21–48.
Victor C., Bowling A., Bond J. and Scambler S. (2003) Loneliness, social isolation and living alone in later life, *GO Findings 17*, Sheffield: Growing Older Programme.
Von Winterfeldt D. and Edwards W. (1986) *Decision Analysis and Behavioural Research*, Cambridge: Cambridge University Press.
Wang W. (1996) Framing effects: dynamics and task domains, *Organisational Behaviour and Human Decision Processes*, 68: 145–57.
Weinstein N. (1983) Reducing unrealistic optimism about illness susceptibility, *Health Psychology*, 2: 11–20.
Weinstein N. (1984) Why it won't happen to me: perceptions of risk factors and susceptibility, *Health Psychology*, 3: 431–57.
Westcott M. (1968) *Towards a Contemporary Psychology of Intuition*, New York: Holt, Rhinehart and Winston.
Wheaton B. (1983) Stress, personal coping resources and psychiatric symptoms: an investigation of interactive models, *Journal of Health and Social Behaviour*, 24: 208–29.
Wilkin D. (1990) Dependency, in Peace S. (ed.) *Researching Social Gerontology*, London: Sage.
Williams S. (1995) Theorising class, health and lifestyles: can Bourdieu help us?, *Sociology of Health and Illness*, 17(5): 577–604.
Willcocks D., Peace S. and Kellaher L. (1987) *Private Lives in Public Spaces*, London: Tavistock.
Wistow G., Waddington E. and Godfrey M. (2003) *Living Well in Later Life: From Prevention to Promotion*, Leeds: Nuffield Institute for Health.
Wright F. (1999) Focus group discussions with older people and care givers. Appendix 2 of *Alternative Models of Care for Older People*, Research Volume 2 of the Royal Commission Report, London: Stationery Office.
Ylanne-Mcewen V. (1999) 'Young at heart': discourses of age identity in travel agency interaction, *Ageing and Society*, 19: 417–40.
Zedner L. (2000) The pursuit of security, in Hope T. and Sparks R. (eds), *Crime, Risk and Insecurity*, London: Routledge.

Author Index

Subject Index